The Prometheus Question

Ex Libr

St. Paul's

RICHMOND, VIRGINIA

D1113272

THE PROMETHEUS QUESTION

A Moral and Theological Perspective on the Energy Crisis

Edited by C.A. Cesaretti

The Seabury Press / New York

3513

1980
The Seabury Press
815 Second Avenue
New York, N.Y. 10017

Copyright © 1980 by C.A. Cesaretti
All rights reserved. No part of this book may be reproduced,
stored in a retrieval system, or transmitted in any form or by
any means, electronic, mechanical, photocopying recording, or
otherwise, without the written permission of The Seabury Press.
Printed in the United States of America.

Library of Congress Cataloging in Publication Data
Main entry under title:
The prometheus question, a moral and theological
   perspective on the energy crisis.
1. Power resources—Moral and religious aspects.
2. Energy policy—Moral and religious aspects.
I. Cesaretti, C.A.

TJ163.2.P77   170   80-19686   ISBN 0-8164-2285-0

In Memoriam

Alfred Johnson

Public Affairs Officer
Executive Council of the Episcopal Church

Contents

Foreword

In 1977 a task force on energy was established by the Executive Council of the Episcopal Church. *The Prometheus Question* is, in part, the result of the research, discussion, and deliberations of this task force. The vignettes which comprise Appendix A witness to the personal search and involvement of every member of the task force.

The task force engaged as their consultant the expertise and vision of the Reverend Scott Paradise. Scott provided the catalyst which forged the many facets of the energy issue and their spokespersons on the task force into a creative pattern of dialogue and growth.

The Reverend Alfred Johnson provided the staff services from The Episcopal Church Center. Al's patient and loving personality was the cement which bound the members of the task force to their task and to each other. His untimely death in the summer of 1979 brought more than sorrow to those who found a strength and refuge in his enormous capacity for compassion, humor, and pastoral concern. *The Prometheus Question* is dedicated to his memory.

The quality of this piece are a witness to the tireless assistance and professionalism of Mrs. Alfreda Barrow and Mr. John Ratti.

> *Charles A. Cesaretti*
> *Public Issues Officer*
> *The Episcopal Church*

Introduction

We are now standing on the threshold of a new world. It could be a world of strife, of hardship, of bitter rivalry over the resources required to maintain our way of life as we have come to know it, or it could be a world in which we grow in awareness of each other's needs and learn to take up our stewardship of earth's riches for our own time and for posterity. It is truly a time of decision in the long history of humankind's existence on earth. We have come increasingly into an understanding of our planet. We have, in a sense, come quite literally into *possession* of it. And yet true responsibility has not come with that possession. But we have arrived at the moment when we must become responsible or lose our world. We know more or less how many souls inhabit this planet and we know what resources are needed to adequately warm, feed, house, and clothe them. But we have also learned that the bounty of the earth is not limitless as we once imagined it to be.

The ancient Greeks knew a great deal about the human race. Their perception reached deep beneath the surface of human motivations. The Greeks knew instinctively what humans were capable of—both the creation of great beauty and the wanton destructiveness inherent in taking without giving. They often saw their gods as acting out, bigger than life, the ways of mortals. The titan Prometheus had the incalculably bad judgement, as far as great Zeus, the father of the gods, was concerned, to be oversympathetic with the pathetic affairs of humans. Zeus believed that the human race would come to a bad end; and although he intervened in their doings from time to time himself, he didn't feel they were really worth the bother. They never seemed to learn the lessons as he set them. But each time Zeus punished the human race or looked as if his patience had run out completely, Prometheus would come to the aid of the humans. His ultimate gift to

1

humankind—for which he paid with everlasting torment—was fire from heaven with which people were to light and warm their world. As Aeschylus described it in his great drama *Prometheus Bound:*

"It was your treasure that he stole, the flowery splendor of all-fashioning fire, and gave to men. . . ."

Zeus, of course, knew that humans were incapable of understanding or using wisely this great gift—which is why he took his terrible revenge on Prometheus. He chained him for all eternity to a rock and had a great eagle tear at his liver during the day and the liver was reconstituted at night so something would be there to tear at the next day. In fact, Zeus did relent after a thousand years and set Prometheus free; but the father of the gods continued to distrust the ways of humans, and Prometheus continued to care for them.

In fact, if we read our newspapers and turn on our television sets today, it would appear that Prometheus was indeed wrong and the father of the gods was correct in his distrust of humans. Westerners, at least, have been so profligate with the fire from heaven that was given as a gratuitous gift that the world as we know it has hovered on the brink of destruction or near-destruction a number of times in the strife between nations to control and *possess* the divine flame of energy.

We live in a world that has grown small. Air travel, telecommunication, all of the advances that have been made which serve to bring the people of the world physically closer together have, in fact, brought us to this ultimate moment of decision. For now one world (and a small word at that), one human family, one table, one hearth are a reality. But the reality we have come to is so simple that, ironically, most people find it hard to grasp.

Christianity speaks of humankind in terms of a family of brothers and sisters, of children with one divine father. But people often listen to the language of religion without hearing what it says.

The energy crisis is not rhetoric; it is real. And although it could indeed give rise to factors which could bring the world as we know it to an end, this energy crisis could also bring with it a blessing in disguise. The energy crisis could finally place us squarely in command of our own destinies. It could force the spoiled children of Prometheus to grow up at last and take *responsible* control of their own world.

Of all of the world's people, the people of the United States have been the most fortunate in their seemingly endless supply of energy sources. But centuries of waste and carelessness are taking their toll. And the desire for more—for more cars, for more electrical appliances, for more heat and more cold—has driven Americans to a seemingly never-ending search for more cheap energy sources to satisfy these monumental needs. We use our own sources and we tap into other nations' energy systems as well in an almost addictive drive to have energy at any price. And we have paid a moral as well as a financial price to satisfy our addiction.

But the picture is not hopeless. We could be standing on the threshold of a new age and not on the brink of disaster. If we listen, if we read the signs carefully, if we hear what our Judaeo-Christian heritage tells us about the charge of humankind to be stewards of God's bounty, we might just come through. It has always been there for us to hear if we would.

When the 1979 Prayer Book of the Episcopal Church was prepared, this growing recognition of human responsibility for our planet was mirrored in many prayers and thanksgivings:

> Almighty God, in giving us dominion over things on earth, you made us fellow workers in your creation; Give us wisdom and reverence so to use the resources of nature, that no one may suffer from our abuse of them, and that generations yet to come may continue to praise you for your bounty . . .

> O God our heavenly Father, you have blessed us and given us dominion over all the earth: Increase our reverence before the mystery of life; and give us new insight into your purposes for the human race, and new wisdom and determination in making provision for its future in accordance with your will . . .

An awareness of the moral and theological dimensions of the energy crisis and of the solutions which must be formulated to cure it has been growing in and out of the traditional religious community. The 1979 General Convention of the Episcopal Church saw the following legislation in response to resolutions about the energy crisis:

> WHEREAS, the growing crisis of energy supply and use is directly and immediately affecting the welfare of each one of us, our nation, and the world; and

WHEREAS, all people are called to be stewards of the gift of God's creation in response to the redeeming love of our Lord, Jesus Christ; and

WHEREAS, the Executive Council, recognizing that concern for human needs places a high priority on an informed Christian community; therefore be it

RESOLVED, the House of Bishops concurring that the 66th General Convention directs the Executive Council to 1) continue the work of the Task Force on Energy and Environment and 2) give a high priority to developing and promoting an educational process to assist the several congregations in assessing and understanding our Christian responsibilities and actions as we are confronted with increasing consumption and dwindling resources throughout the world.

And on January 10, 1980, President Carter met with a group of religious leaders at the White House to spell out his own perception of the moral and theological concerns raised by the crisis. The remarks which follow formed the key portions of an address which President Carter delivered to the assembled leaders:

. . . When I was at Camp David last July I had a group of people who are very important to me come and see me from one day to another to give me advice on what our nation should do with this energy crisis and at that time an existing serious challenge. One of the groups was comprised of some of you. A quiet, meditative period of intense, unrestrained frank discussion of the moral and ethical and religious principles that were involved in meeting the crisis of energy.

It might seem strange to some, not to you, that the conservation of oil has a religious connotation but when God created the earth and gave human beings dominion over it, it was with the understanding on the part of us, then and down through the generations, that we are indeed stewards under God's guidance. To protect not only those who are fortunate enough to grasp an advantage or a temporary material blessing or enjoyment but to husband those bases for enjoyment and for a quality of life for those less fortunate in our own generation, especially for those who will come after us.

Our country is comprised of profligate wasters of the earth's precious resources, not because of an innate selfishness, but because we have been overly endowed by God with those material

blessings. We have seldom experienced limits on our lives because of a withholding the production of food or fiber or building
materials or energy itself, access to warm oceans, wonderful
climates, rich land. God has given us these things.

But lately, in the last few years, and particularly in the last few
months, we have begun to see that we not only have a responsibility to now and future Americans, but also to those who live on
earth now and will live in the future.

No one could anticipate the broad use of petroleum products.
A few years ago it was looked upon as a rapidly expendable but
inexhaustible supply of just fuel, to burn, to make heat. Petroleum products now are used to make food, to make medicine,
and for other uses that directly affect the quality of life of human
beings, in addition to the burning of the fuel for heat or propulsion.

We have seen also the interrelationship between energy supplies and peace; between energy supplies and the protection of
religious beliefs. The right of people to be free is directly tied to
adequate supplies of energy in the modern, fast-changing technological world. I am not a theologian, I don't understand all of the
relationships between these subjects, but I am particularly grate-
ful that you as religious leaders have come to the White House to
explore not only the theoretical, theological aspects of stewardship and conservation; but also in more depth, how you as religious leaders, and others like you in every church and syna-
gogue throughout our country, might explore even further the
aspects of living in accordance with God's will, to promote the
concepts of peace and freedom and unselfishness and humility
and responsibility for the well-being of others.

I am determined that our nation will be strong. I am determined that our nation will stay free. I am determined that our nation will hold high the banner of human rights for ourselves and
for others. And I am determined that the American people, as
best we can, will be educated about those interrelationships that
are so important to us all.

You can help greatly with this concept, because still in our
blessed land many people cannot accept easily the concept of material limits. There are only two ways to resolve the energy problem in the future. One is to produce more energy in our own
country, preferably with replenishable supplies, where the origin
is the sun, or through flowing water, or prevailing winds, we
might derive energy without limit.

Another source of traditional energy that is not so plentiful is
petroleum products. The production of more energy is one basic
approach, and there is only one other. That is the conservation of
energy in all forms; the elimination of waste. And along with that

is the better sharing of energy among all of us for the well-being of our country and the individuals who live here.

I don't think that either one of these programs or concepts or commitments needs to cause a deterioration in the quality of life of our people. It is not a sacrifice to eliminate waste. It can be a blessing, not necessarily in disguise, to eliminate a dependence of one person riding in a very heavy, very expensive, very wasteful vehicle. It is not a contributing factor to the quality of life to have a home that requires twice as much energy to heat it as is necessary, or to have little clothing on in a home when a few degrees in temperature lower and a sweater should let us realize that there is a change of seasons outside, that God's plan is still working on an annual basis.

And I think the drawing together of families to discuss this challenge, which is becoming ever more important in the lives of human beings, to discuss how we will meet this in our own personal life, can be a coalescing factor, strengthening family ties and therefore the community. And those, of course, are the basic elements of American life. I think we will turn more to the simple things of life: quiet discussions at home; the sharing of experiences; a walk in the woods; a look at God's earth; and the elimination of the frantic dash from one place to other, where we lose sight of what we are seeking at the end of that trip or that dash, where our senses are pretty well desensitized, as we move through the views that God has given us.

I am not trying to preach a sermon to you, but I am very deeply concerned about how Americans look upon resolving the energy question. It will require unity. It will require some sacrifice. It will require courage. It will require persistence or tenacity. It will require knowledge. It will require the assessment of the priorities that we have established in our lives to measure what is a good life, and the quality of life might be. . . .

The Prometheus Question is a resource book for adults seeking to understand the energy crisis in terms of their lives and their faith. Because the issues raised by the crisis are tied to almost every facet of our daily lives, the four discussion sessions outlined in this book take their impetus from many sources. Two appendices of readings are provided at the back of the book. Appendix A contains first-hand, first-person accounts of the encounters of people with the "facts of life" of the crisis; their experiences are both universal and very personal. Each of the accounts in Appendix A concludes with a useful group of appropriate study or thought questions about the reading. Appendix B is a selection of articles from national magazines and news-papers on the crisis. They provide scientific background and other data about alternate energy sources. These articles are included not only to provide data and perspective but to provide the opportunity to establish a process of approaching the secular media from a Biblical, theological, and moral dimension.

But the main source for these discussions is the participant and his or her experience and awareness. The sessions are grounded in Scripture as well as in topical readings. They are intended to help the participants identify their roles in the crisis, their responsibility for curing it. The sessions are designed to snowball—insights and understandings growing as they go.

The crisis and the solution rest with us all.

Notes on Organizing the Group

The Prometheus Question is a resource book for adults. The sessions are informal, but as there is a variety of material provided for the discussions in the book and a wealth of material the participants will encounter in their everyday lives and bring to the group sessions, it would be wise for there to be established leadership—either an appointed or elected leader for the whole series or a rotating leadership with a leader elected for each session by the participants. A certain amount of the material is controversial—as are the issues they address—and a certain attempt will have to be made to keep the sessions orderly enough for the participants to profit from them.

The ideal length for these sessions would be 1½ to 2 hours. They could be scheduled after a Sunday morning church service in the traditional adult education slot or as an evening series. They may be used at a weekend parish conference or by a cluster of parishes in your area or by an ecumenical study group.

This is not a series to be dropped into and out of, despite its relative informality. The participants should be committed to learning something about themselves and the world they live in. They will be asked to think a lot and to do some amount of demanding homework. If your church is accustomed to having persons entering into the serious discussion groups "contract" with the leader or with each other, sealing their commitment to the program, this would be a good occasion for doing it.

Make the series known to your congregation well in advance of the sessions. Have copies of the volume available in the church office or some other centrally located place to be looked over before the commitments are made. It would be best to have a pre-series signing up session, short if necessary, a week prior to the beginning of the series. At this point, the leader should suggest that everyone read the Introduction in advance of the first meeting, with special emphasis on the excerpts from President Carter's address to religious leaders, which are included in the Introduction. This address sets the tone, in a sense, for the whole series.

The box which follows will be helpful to you as you make your plans to get your sessions moving. Of course, it is a suggested plan of procedure which should be adjusted to the special needs of your group.

GETTING STARTED

Suggestions for the Person or Persons
Who Will Lead the Sessions

—Announce the series in your parish bulletin if there is one. If not, put a clearly printed notice on the most frequently looked at parish bulletin board.
—Provide sign up opportunities. If you have a parish bulletin you could have a coupon attached to it that could be cut out, filled in, and dropped off at the parish office or mailed.
—Secure necessary resource materials that might help you, the leader, in launching and carrying through the series. Your parish library or your local public library may have useful materials on adult education experiences. A particularly helpful book to use in preparing this kind of experience has been published as part of The Church's Teaching Series. It is called *Equipping God's People: Basic Concepts for Adult Education* (Evans and Hayes, The Seabury Press, 1979).
—Once you know how many people will be in your group, purchase books for each participant.
—You should read all of *The Prometheus Question* before you meet with the group for the first time.

—Plan a pre-series session to:
  —introduce the series and the issues it raises
  —suggest that everyone read the introduction with special emphasis on President Carter's address
  —choose a facilitator for the first session
  —announce the time, date, and place of meeting
  —discuss the concept of keeping a logbook or some other record of their energy use: this record would be kept throughout the series
  —decide on the most convenient time for everyone to meet and pass out contracts if you have chosen to use them

Session 1

Bible Study and Discussion

Begin the session with the Bible. All of the sessions will begin with a reading from Scripture. The leader or a member of the group should begin the first session by reading aloud Psalm 104. It is beautiful and provides a moving way to begin thinking about Creation and Stewardship.

104

Bless the LORD, my soul:
O Lord my God, thou art great indeed,
clothed in majesty and splendour,
and wrapped in a robe of light.
Thou hast spread out the heavens like a tent
and on their waters laid the beams of thy pavilion;
who takest the clouds for thy chariot,
riding on the wings of the wind;
who makest the winds thy messengers
and flames of fire thy servants;
thou didst fix the earth on its foundation
so that it never can be shaken;
the deep overspread it like a cloak,
and the waters lay above the mountains,
At thy rebuke they ran,
at the sound of thy thunder they rushed away,
flowing over the hills,
pouring down into the valleys
to the place appointed for them.
Thou didst fix a boundary which they might not pass;
they shall not return to cover the earth.

Thou dost make springs break out in the gullies,
so that their water runs between the hills

The wild beasts all drink from them
the wild asses quench their thirst;
the birds of the air nest on their banks
and sing among the leaves
From thy high pavilion thou dost water the hills;
the earth is enriched by thy provision.
Thou makest grass grow for the cattle
    and green things for those who toil for man,
bringing bread out of the earth
and wine to gladden men's hearts,
oil to make their faces shine
and bread to sustain their strength.
    The trees of the Lord are green and leafy,
the cedars of Lebanon which he planted;
the birds build their nest in them,
the stork makes her home in their tops.
High hills are the haunt of the mountain-goat,
and boulders a refuge for the rock badger.

Thou hast made the moon to measure the year
and taught the sun where to set.
When thou makest darkness and it is night,
all the beasts of the forest come forth;
the young lions roar for prey,
seeking their food from God.
When thou makest the sun rise, they slink away
    and go to rest in their lairs;
but man comes out to his work
    and to his labours until evening.
Countless are the things thou hast
so made, O Lord.
Thou hast made all by thy wisdom;
and the earth is full of thy creatures,
beasts great and small.

Here is the great immeasurable sea,
in which move creatures beyond number
Here ships sail to and fro,
here is Leviathan whom thou hast made thy plaything.

All of them look expectantly to thee
to give them their food at the proper time;
what thou givest them they gather up;
when thou openest thy hand, they eat their fill.
Then thou hidest thy face, and they are restless and troubled;

when thou takest away their breath they fail
[and they return to the dust from which they came];
but when thou breathest into them, they recover;
  thou givest new life to the earth.

May the glory of the LORD stand for ever
and may he rejoice in his works!
When he looks at the earth, it quakes;
when he touches the hills, they pour forth smoke.

I will sing to the LORD as long as I live,
all my life I will sing psalms to my God.
May my meditation please the LORD,
as I show my joy in him!
Away with all sinners from the earth
and may the wicked be no more!

Bless the LORD, my soul.

O praise the LORD.

Talk about Psalm 104. These questions may help the group.
—According to this psalm, who owns the earth and its resources?
—For whose benefit was it created?
—What role do humans play in the whole order of things in the universe?
—Is this psalm just beautiful poetry, or does it have something to say to you about the production, conservation, and proper use of energy?
—What *does* the word "stewardship" mean to you?
—Do you really believe that God created the complex modern world as we know it, or did he just create the world of the Biblical herdsmen, fishermen, and small farmers?
—How does the psalm parallel the creation narrative in the book of Genesis? A member of the group might want to read this narrative aloud at this point.
—Compare the psalm with Robert Grant's 1833 adaptation of it for the hymn *O Worship the King* (Hymn 288 in the 1940 hymnal of The Episcopal Church):

O worship the King, all glorious above!
O gratefully sing his power and his love!
Our shield and defender, the Ancient of Days,
Pavilioned in splendor, and girded with praise.

O tell of his might! O sing of his grace!
Whose robe is the light, whose canopy space.
His chariots of wrath the deep thunderclouds form,
And dark is his path on the wings of the storm.

The earth, with its store of wonders untold,
Almighty, thy power hath founded of old,
Hath stablished it fast by a changeless decree,
And round it hath cast, like a mantle, the sea.

Thy bountiful care, what tongue can recite?
It breathes in the air; it shines in the light;
It streams from the hills; it descends to the plain,
And sweetly distils in the dew and the rain.

—Try recasting the psalm, as Robert Grant did, in your own words. If some group members are interested in doing more than paraphrasing the psalm, suggest that they work on it over the next week and share the results the next time.

Moving Along

The group has been asked to read President Carter's address which is included in the Introduction to this book.
—What do you think of the address?
—Do you think it is "just politics" or do you find some true meaning and insight in it?
—Members of the group may choose to read aloud and discuss the paragraph of the address they find has most meaning for them. Or perhaps some group members would like to read and discuss the section with which they most disagree.
—What does President Carter mean when he says: ". . . We will turn more to the simple things of life"?

One of the thrusts of President Carter's address was the suggestion that modern people in America are apt to become desensitized to the world around them because they move through it, in effect, so rapidly. By extension it might seem likely that being desensitized or even unaware of what we do and how we expend energy in every sense of the word might inevitably lead us to waste and to loss. The group might well discuss or begin discussing how they use energy in their lives.
—How did you get here for this meeting? How will you return home?

If you drive a car, how many miles do you drive in a week? How many miles does it "get to the gallon"?

—How is your house heated? How else do you use energy in your house? How many appliances do you have that require electricity?

—Is there any appliance or energy-using device in your house that you could do without easily?

—Do you know how to read your gas and/or electric meter? When you go home record the figures on your meters on a piece of paper and then record the reading on the same meters just before you start out to the next session of this series.

—How many light bulbs do you have in your house? What is their total wattage?

Have the group take the test on the following page. It is intended to assess how much one knows about energy. You could read the questions aloud.

Energy Conservation Quiz
TESTING YOUR ENERGY QUOTIENT

Rating: 15–12 "High" E.Q.; 11–9 Average; 8–0 Low.
Score: One point for each correct answer, and don't peek ahead.

QUESTION 1: Keeping house heating temperatures in the winter below 75 degrees generally results in what percent of energy savings per degree? A. None B. 1–3 percent C. 3–10 percent D. 10–15 percent E. More than 15 percent
ANSWER: C. About 3–10 percent per degree based on statistics from the National Bureau of Standards. In summer, each degree above 75 saves about 3–7 percent depending on geographic location.

QUESTION 2: Setting thermostats lower at night in the winter when you go to bed causes you to use as much fuel restoring the heat level in the morning as you save during the setback period. True False
ANSWER: False. A computer study by Honeywell showed 5 to 16 percent savings beyond the cost of fuel recovery with either manual or clock-thermostat night setback. By delaying starts of the heat recovery period in the morning from 6 to 8 a.m. you save an additional 4 percent of fuel by using the morning sun's natural heating of the home.

QUESTION 3: The best way to control steam or hot-water radiatior heat is by screwing the regulator knob up or down, or by opening a window slightly. True False
ANSWER: False. The regulator knob is imprecise. Opening a window overcompensates, causes the radiator to put out more heat, and wastes already heated air by dumping it out the window. Use of a thermostatic valve in place of the standard screw-knob can cut energy requirements as much as 27 percent.

QUESTION 4: An electronic air cleaner, installed in the duct-work of a forced air system to filter out airborne dust and pollen, is one of the heaviest energy-using appliances in the home. True False
ANSWER: False. Typical electronic air cleaners use about as much energy as a lightbulb and only operate when the forced air system is on. In commercial buildings they reduce to a minimum the requirement for outside air ventilation and thus cut heating or cooling needs. Even if you have only a furnace filter, clean it regularly so the furnace doesn't have to work harder to get heat to more remote areas. An inexpensive device called a filter flag can be installed to tell when the air flow is clogged and the filter should be cleaned or replaced.

QUESTION 5: A shower uses more water than a bath. True False
ANSWER: False. A shower can use as much as five gallons less water than a bath, according to Money Magazine, October 1973.

QUESTION 6: Outside air can be used in the summer to cool a house. True False
ANSWER: True. After the sun goes down, cool evening air can be let in to help cooling systems. The homeowner can open a window or install an "economizer" system that automatically monitors outside temperatures and opens a damper to let the air in when it reaches the desired temperature for indoor air.

QUESTION 7: Winter heating accounts for more than half the energy used in the typical home, whether it comes from oil, gas, electricity or coal. True False
ANSWER: True. Attributed to the Tennesssee Valley Authority by Fueloil & Oil Heat Magazine, January 1974.

QUESTION 8: Humidity levels in homes affect: A. Static in Air B. Respiration C. Heating or Cooling Needs D. All of these E. None of these
ANSWER: D. All of these.

QUESTION 9: Warm-air leakage or cold-air infiltration in winter due to lack of caulking or adequate weatherstripping can amount to no more than a 10 percent increase in heating bills. True False
ANSWER: False. It can go as high as 15 to 30 percent according to a report from the Office of Consumer Affairs, 1973.

QUESTION 10: Storm or double-pane windows can cut heat loss through window glass: A. 10 percent B. 30 percent C. 50 percent
ANSWER: C. According to a 1973 report from the Office of Consumer Affairs.

QUESTION 11: A leaky hot-water faucet that fills an ordinary cup in 10 minutes wastes heated water at the rate of more than 3,000 gallons a year. True False
ANSWER: True. According to a 1973 report from the Office of Consumer Affairs.

QUESTION 12: A one-story home has less heat-loss than a two-story home of the same heat space. True False
ANSWER: False.

QUESTION 13: The water heater is the second largest energy user in the home. True False

ANSWER: True. It uses 13 percent of home energy according to Homebuilding Business, March 1974. Tennessee Valley Authority figures confirm this and list house heating as highest (27 percent) followed by water heater and then air conditioning (12.5 percent).

QUESTION 14: A temperature setting of about 150 degrees at the middle of the water heater thermostat dial is required for adequate bathing and for disinfecting clothes or dishes. True False
ANSWER: False. A lower setting of 120 degrees is sufficient for bathing, washing clothes or dishes and will save from 20 to 25 percent of energy required for the water heater, according to Homebuilding Business, March 1974. (To disinfect clothes and dishes, a temperature of 180 degrees for two minutes is required.)

QUESTION 15: Light bulbs produce nearly four times as much light per watts as fluorescent lamps.
ANSWER: False. Fluorescents produce nearly four times as much as general service light bulbs according to Homebuilding Business, March 1974.

After the examination has been completed read the correct answers and let the group mark their own papers.

Preparing for the Next Session
Start to discuss what the agenda for the next session ought to be.
—Suggest that each person begin keeping for the remainder of the commitment a daily log in a small pocket notebook of every use of energy in the course of the day—from waking to sleeping.
—Suggest that each person keep in the same logbook a notation of everything they do in the course of a day that does not require the use of a machine or other energy-using device.
—Assign each person the responsibility of choosing one article from each of the Appendices—A and B—for reading, they should be prepared to talk about them the next time. There will be some duplications—but that can be sorted out in the second session. Remember to take a look at the study questions which follow the accounts in Appendix A.
—If there is time, each person in the group should try to find a passage in the Bible involving the relationship of God to humankind and humankind to the natural world. Encourage people who have newer translations of Scripture to use them, as fresh language will help the group to see things in a new light.

Ending

Read aloud the words of institution and consecration in the Eucharist—either from the Book of Common Prayer or another Eucharistic rite.

—What do these words mean to you? Are they important to you?

—What do they tell you about the involvement of the Divine in the human world?

—Do you feel that involvement? If you do, how? If you don't, why?

A great deal of getting acquainted with the program and each other has taken place in this session. It would probably be wise to end it simply. The *General Thanksgiving* on page 836 of the Prayer Book would be a good choice for ending, or one of the two Thanksgivings for the Natural Order on page 840 of the Prayer Book. Have a volunteer read one or more of these thanksgivings.

And at the Very End

Have the group assess the learnings they have gained in the session.

—What did you learn?

—Identify any new insights you gained.

—What issue discussed in the session would you like to know more about?

Session 2

Bible Study and Discussion

Begin the session with a Bible reading. Since the group has begun to look at their own use of energy in preparation for this session they will also begin to look at how they fit into the greater community in which they live and perhaps to consider some of the obligations that position involves. The reading is from Deuteronomy 24:14–22.

> You shall not keep back the wages of a man who is poor and needy, whether a fellow-countryman or an alien living in your country in one of your settlements. Pay him his wages on the same day before sunset, for he is poor and his heart is set on them: he may appeal to the LORD against you, and you will be guilty of sin.
>
> Fathers shall not be put to death for their children, nor children for their fathers; a man shall be put to death only for his own sin.
>
> You shall not deprive aliens and orphans of justice nor take a widow's cloak in pledge. Remember that you were slaves in Egypt and the LORD your God redeemed you from there; that is why I command you to do this.
>
> When you reap the harvest in your field and forget a swathe, do not go back to pick it up; it shall be left for the alien, the orphan, and the widow, in order that the LORD your God may bless you in all that you undertake.
>
> When you beat your olive-trees, do not strip them afterwards; what is left shall be for the alien, the orphan, and the widow.
>
> When you gather the grapes from your vineyard, do not glean afterwards; what is left shall be for the alien, the orphan, and the widow. Remember that you were slaves in Egypt; that is why I command you to do this.

Like many passages in Scripture, this is a moving call for justice for the poor, the needy, and the servant, the fatherless, the widow, and the sojourner.

—Why do you suppose this passage like much of Scripture expresses more concern about justice for the poor than for the rich?

—Are you involved in any activity in your community that attempts to help a person or persons with specific kinds of problems? If so, what is the involvement? If you have no such involvement, is it a conscious decision on your part not to have it, or do you simply not have the time in the particular life-style you're a part of?

—Do you feel responsible and/or involved with any person or persons you come in contact with outside your immediate family?

—If you met a beggar or a panhandler in the street, would you be apt to give him or her something? If you do, why do you do it? If you don't, why don't you?

—Do you think what any single individual does or does not do will or can help the community in which they live in any important way?

—Can you recall a passage from the Gospels that echoes the concerns of Deuteronomy?

Ask for any passage of Scripture to be read on the relationship of God to humankind and humankind to the natural world. This was one of the assignments from the previous session.

—What does the passage mean to the person who found it?

—What does it mean to the group?

Moving Along

Have the group share the selections from the Appendices which they chose to read. There will probably be time for selections from the articles to read aloud but not whole articles. The essence of the article should be shared with the group.

Have the group share their lists of energy-using and non-energy-using activities, and their experiences with meter reading. Some lists could be read; others could be written on newsprint.

—What did keeping these logs or lists tell you about yourself?

—Were you surprised at what you found?

—What similarities does the group spot in sharing their lists? What dissimilarities?

What do you know about the energy resources of the community you live in? It is reasonably likely that the group assembled will all be from the same general geographical area.

—How is electricity generated in your town?

—Is natural or bottled gas used?
—Is your town or community heavily "motorized"? Do you drive a car yourself? Do you have more than one car in the family?
—Is there industry in your town? What kind and what use of energy sources does it make? Do you live in a rural community? What motor operated equipment is used in local farming?

Preparing for the Next Session

Start to discuss moving forward for next week. People are probably beginning to spot areas of special interest in their local communities and in the Appendix articles they are reading.
—Suggest that the group begin clipping the energy oriented articles from their local newspapers and magazines and from national magazines and newspapers as well to share with the group as a whole in the third session.
—Suggest that for session 3 each person write their own personal anecdote about their emerging understanding of the uses of energy and the energy crisis. Make sure they look at the material in Appendix A once more. The vignettes there could well serve as models.
—Assign each person the responsibility of choosing an article from each of the appendices, which they have not previously read. However, this time specify that it be, if possible, on a topic (in the case of Appendix B especially) that they know *little* about. And in this case suggest that after reading another personal experience from Appendix A, the group member read the study questions which follow each of the selections, and then make up a set of study questions of his or her own.
—Have the group start looking for a local energy problem or issue in the community that especially interests them, with an eye toward finding out more about it.
—If there is time, have group members spot passages in the Bible about waste or squandering of life and/or energy.

Ending

In Matthew 15:32–39, read the story aloud of the loaves and fishes.
—What is the miracle in this story? Was the story of Christ feeding the multitude told to show what miraculous powers he had? Or was the miracle a more complicated one than that?

—Is there *anything* we can share with others that *matters?*
—Can you see a connection between this story of the loaves and fishes and last week's look at the Eucharist?
—What does *sharing* mean?

Instead of reading set prayers to end the session, have a few moments of silence and then suggest that, simply and in one sentence, each group member expresses gratitude for something or someone in his or her life.

And at the Very End

Have the group assess the learnings they have gained in the session.
—What did you learn?
—Identify any new insights you gained.
—What issue discussed in the session would you like to know more about?

Session 3

Bible Study and Discussion

Begin with the rather extensive readings which follow. These readings from the Bible have to do with human fallibility. The group has been involved with personal record-keeping and inventory which has probably given them some idea of their own fallibility, and areas for reflection and change. In this session they will begin to ask some very searching questions about their community. They are going to find that there are a number of clay feet among the public officials and others in their community. Scripture can be quite frank and unflinching in the area of human fallibility and these rather dramatic passages could help the group come to terms with what they will, inevitably, find. If time permits, the material which follows could well be acted out. But it certainly should certainly be read aloud by different group members, in any event.

These readings refer to incidents in the lives of four biblical heroes—Noah, Saul, and David of the Old Testament and Peter of the New. As your read the passages consider what they say about human nature.

The story of Noah is first.

> When the LORD saw that man had done much evil on earth and that his thoughts and inclinations were always evil, he was sorry that he had made man on earth, and he was grieved at heart. He said, 'This race of men whom I have created, I will wipe them off the face of the earth—man and beast, reptiles and birds. I am sorry that I ever made them.' But Noah had won the LORD's favour.
>
> This is the story of Noah. Noah was a righteous man, the one blameless man of his time; he walked with God.
>
> *Genesis 6:5–10*

23

Noah alone of the whole human race God felt worthy to survive.
Then follows the story of the ark. After the flood the story continues.

> Noah, a man of the soil, began the planting of vineyards. He
> drank some of the wine, became drunk and lay naked inside his
> tent. When Ham, father of Canaan, saw his father naked, he told
> his two brothers outside. So Shem and Japheth took a cloak, put
> it on their shoulders and walked backwards, and so covered their
> father's naked body; their faces were turned the other way, so
> that they did not see their father naked. When Noah woke from
> his drunken sleep, he learnt what his youngest son had done to
> him, and said:

> 'Cursed be Canaan,
> slave of slaves
> shall he be to his brothers.'

> And he continued:

> 'Bless, O LORD,
> the tents of Shem;[b]
> may Canaan be his slave.
> May God extend[c] Japheth's bounds,
>
>   let him dwell in the tents of Shem,
> may Canaan be their slave.'

> After the flood Noah lived for three hundred and fifty years, and
> he was nine hundred and fifty years old when he died.
> *Genesis 9:20–29*

Saul also was chosen by God. He was to deliver God's people.

9

> There was a man from the district of Benjamin, whose name
> was Kish son of Abiel, son of Zeror, son of Bechorath, son of
> Aphiah a Benjamite. He was a man of substance, and had a son
> named Saul, a young man in his prime; there was no better man
> among the Israelites than he. He was a head taller than any of his
> fellows.

10

> Samuel took a flask of oil and poured it over Saul's head, and
> he kissed him and said, 'The LORD anoints you prince over his

people Israel; you shall rule the people of the LORD and deliver them from the enemies round about them. You shall have a sign that the LORD has anointed you prince to govern his inheritance:
1 Samuel 9:1–2; 10:1

He successfully led Israel's forces against their enemies on every side including the Amalekites.

Then the word of the LORD came to Samuel: 'I repent of having made Saul king, for he has turned his back on me and has not obeyed my commands.' Samuel was angry; all night he cried aloud to the LORD. Early next morning he went to meet Saul, but was told that he had gone to Carmel; Saul had set up a monument for himself there, and had then turned and gone down to Gilgal. There Samuel found him, and Saul greeted him with the words, 'The LORD's blessing upon you! I have obeyed the LORD's commands.' But Samuel said, 'What then is this bleating of sheep in my ears? Why do I hear the lowing of cattle?' Saul answered, 'The people have taken them from the Amalekites. These are what they spared, the best of the sheep and cattle, to sacrifice to the LORD your God. The rest we completely destroyed.' Samuel said to Saul, 'Let be, and I will tell you what the LORD said to me last night.' 'Tell me', said Saul. So Samuel went on, 'Time was when you thought little of yourself, but now you are head of the tribes of Israel, and the LORD has anointed you king over Israel. The LORD sent you with strict instructions to destroy that wicked nation, the Amalekites; you were to fight against them until you had wiped them out. Why then did you not obey the LORD? Why did you pounce upon the spoil and do what was wrong in the eyes of the LORD?' Saul answered Samuel, 'But I did obey the LORD; I went where the LORD sent me, and I have brought back Agag king of the Amalekites. The rest of them I destroyed. Out of the spoil the people took sheep and oxen, the choicest of the animals laid under ban, to sacrifice to the LORD your God at Gilgal.' Samuel then said:

Does the LORD desire offerings and sacrifices
as he desires obedience?
Obedience is better than sacrifice,
and to listen to him than the fat of rams.
Defiance of him is sinful as witchcraft,
yielding to men*[g]* as evil as*[h]* idolatry.*[i]*
Because you have rejected the word of the LORD,
the LORD has rejected you as king.

Saul said to Samuel, 'I have sinned. I have ignored the LORD's
command and your orders: I was afraid of the people and de-
ferred to them. But now forgive my sin, I implore you, and come
back with me, and I will make my submission before the LORD.'
Samuel answered, 'I will not come back with you; you have
rejected the word of the LORD and therefore the LORD has rejected you as king over Israel.' He turned to go, but Saul caught
the edge of his cloak and it tore. And Samuel said to him, 'The
LORD has torn the kingdom of Israel from your hand today and
will give it to another, a better man

I Samuel 15:10–28

David succeeded Saul. He became Israel's greatest king and led
Israel on a succession of military victories until:

11

AT THE TURN OF THE YEAR, WHEN KINGS take the field, David
sent Joab out with his other officers and all the Israelite forces,
and they ravaged Ammon and laid siege to Rabbah, while David
remained in Jerusalem. One evening David got up from his couch
and, as he walked about on the roof of the palace, he saw from
there a woman bathing, and she was very beautiful. He sent to
inquire who she was, and the answer came, 'It must be Bath-
sheba daughter of Eliam and wife of Uriah the Hittite.' So he
sent messengers to fetch her, and when she came to him, he had
intercourse with her, though she was still being purified after her
period, and then she went home. She conceived, and sent word
to David that she was pregnant. David ordered Joab to send
Uriah the Hittite to him. So Joab sent him to David, and when he
arrived, David asked him for news of Joab and the troops and
how the campaign was going; and then said to him, 'Go down to
your house and wash your feet after your journey.' As he left the
palace, a present from the king followed him. But Uriah did not
return to his house; he lay down by the palace gate with the
king's slaves. David heard that Uriah had not gone home, and
said to him, 'You have had a long journey, why did you not go
home?' Uriah answered David, 'Israel and Judah are under can-
vas,[b] and so is the Ark, and my lord Joab and your majesty's
officers are camping in the open; how can I go home to eat and
drink and to sleep with my wife? By your life, I cannot do this!'
David then said to Uriah, 'Stay here another day, and tomorrow I
will let you go.' So Uriah stayed in Jerusalem that day. The next
day David invited him to eat and drink with him and made him

drunk. But in the evening Uriah went out to lie down in his blanket[c] among the king's slaves and did not go home.

The following morning David wrote a letter to Joab and sent Uriah with it. He wrote in the letter, 'Put Uriah opposite the enemy where the fighting is fiercest and then fall back, and leave him to meet his death.' Joab had been watching the city, and he stationed Uriah at a point where he knew they would put up a stout fight. The men of the city sallied out and engaged Joab, and some of David's guards fell; Uriah the Hittite was also killed. Joab sent David a dispatch with all the news of the battle and gave the messenger these instructions: 'When you have finished your report to the king, if he is angry and asks, "Why did you go so near the city during the fight? You must have known there would be shooting from the wall. Remember who killed Abimelech son of Jerubbesheth. It was a woman who threw down an upper millstone on to him from the wall of Thebez and killed him! Why did you go so near the wall?"—if he asks this, then tell him, "Your servant Uriah the Hittite also is dead." '

So the messenger set out and, when he came to David, he made his report as Joab had instructed. David was angry with Joab and said to the messenger, 'Why did you go so near the city during the fight? You must have known you would be struck down from the wall. Remember who killed Abimelech son of Jerubbesheth. Was it not a woman who threw down an upper millstone on to him from the wall of Thebez and killed him? Why did you go near the wall?' He answered, 'The enemy massed against us and sallied out into the open; we pressed them back as far as the gateway. There the archers shot down at us from the wall and some of your majesty's men fell; and your servant Uriah the Hittite is dead.' David said to the man, 'Give Joab this message: "Do not let this distress you—there is no knowing where the sword will strike; press home your attack on the city, and you will take it and raze it to the ground"; and tell him to take heart.'

When Uriah's wife heard that her husband was dead, she mourned for him; and when the period of mourning was over, David sent for her and brought her into his house. She became his wife and bore him a son. But what David had done was wrong in the eyes of the LORD.

2 Samuel 11

In the New Testament Peter became pre-eminent among the apostles. He first confessed Jesus to be the Christ. The crisis came.

In the evening he came to the house with the Twelve. As they sat at supper Jesus said, 'I tell you this: one of you will betray me—one who is eating with me.' At this they were dismayed; and one by one they said to him, 'Not I, surely?' 'It is one of the Twelve', he said, 'who is dipping into the same bowl with me. The Son of Man is going the way appointed for him in the scriptures; but alas for that man by whom the Son of Man is betrayed! It would be better for that man if he had never been born.'

During supper he took bread, and having said the blessing he broke it and gave it to them, with the words: 'Take this; this is my body.' Then he took a cup, and having offered thanks to God he gave it to them; and they all drank from it. And he said, 'This is my blood, the blood of the covenant, shed for many. I tell you this: never again shall I drink from the fruit of the vine until that day when I drink it new in the kingdom of God.'

After singing the Passover Hymn, they went out to the Mount of Olives. And Jesus said, 'You will all fall from your faith; for it stands written: "I will strike the shepherd down and the sheep will be scattered." Nevertheless, after I am raised again I will go on before you into Galilee.' Peter answered, 'Everyone else may fall away, but I will not.' Jesus said, 'I tell you this: today, this very night, before the cock crows twice, you yourself will disown me three times.' But he insisted and repeated: 'Even if I must die with you, I will never disown you.' And they all said the same.

WHEN THEY REACHED A PLACE CALLED Gethsemane, he said to his disciples, 'Sit here while I pray.' And he took Peter and James and John with him. Horror and dismay came over him, and he said to them, 'My heart is ready to break with grief; stop here, and stay awake.' Then he went forward a little, threw himself on the ground, and prayed that, if it were possible, this hour might pass him by. 'Abba, Father,' he said, 'all things are possible to thee; take this cup away from me. Yet not what I will, but what thou wilt.'

He came back and found them asleep; and he said to Peter, 'Asleep, Simon? Were you not able to stay awake for one hour? Stay awake, all of you; and pray that you may be spared the test. The spirit is willing, but the flesh is weak.' Once more he went away and prayed.[k] On his return he found them asleep again, for their eyes were heavy; and they did not know how to answer him.

The third time he came and said to them, 'Still sleeping? Still taking your ease? Enough![l] The hour has come. The Son of Man is betrayed to sinful men. Up, let us go forward! My betrayer is upon us.'

Suddenly, while he was still speaking, Judas, one of the Twelve, appeared, and with him was a crowd armed with swords and cudgels, sent by the chief priests, lawyers, and elders. Now the traitor had agreed with them upon a signal: 'The one I kiss is your man; seize him and get him safely away.' When he reached the spot, he stepped forward at once and said to Jesus, 'Rabbi', and kissed him. Then they seized him and held him fast.

One of the party[m] drew his sword, and struck at the High Priest's servant, cutting off his ear. Then Jesus spoke: 'Do you take me for a bandit, that you have come out with swords and cudgels to arrest me? Day after day I was within your reach as I taught in the temple, and you did not lay hands on me. But let the scriptures be fulfilled.' Then the disciples all deserted him and ran away.

Among those following was a young man with nothing on but a linen cloth. They tried to seize him; but he slipped out of the linen cloth and ran away naked.

Mark 14:17–31; 43–50; 49–52

They arrested Jesus and took him before the high priest for questioning.

Meanwhile Peter was still below in the courtyard. One of the High Priest's serving-maids came by and saw him there warming himself. She looked into his face and said, 'You were there too, with this man from Nazareth, this Jesus.' But he denied it: 'I know nothing,' he said; 'I do not understand what you mean.' Then he went outside into the porch;[p] and the maid saw him there again and began to say to the bystanders, 'He is one of them'; and again he denied it.

Again, a little later, the bystanders said to Peter, 'Surely you are one of them. You must be; you are a Galilean.' At this he broke out into curses, and with an oath he said, 'I do not know this man you speak of.' Then the cock crew a second time; and Peter remembered how Jesus had said to him, 'Before the cock crows twice you will disown me three times.' And he burst into tears.

Mark 14:66–72

—-What do these four figures have in common? Are they different from other Biblical characters you have encountered?

—Are they at all like people you know or have known about?

—Is it possible to plan a new approach to energy and energy conservation that would avoid the problems that are connected with human fallibility?

—From what you have learned so far, is there any agency or company in your community involved with energy that is doing things properly? If so, do you see any way the methods they are using could be tied into other energy systems in your community?

—Would you, if you had training in engineering, be prepared to tackle the energy problems of your community? If you think so, what would you do first?

Moving Along

In the last session, it was suggested that members write their own personal anecdotes about their encounters with the use of energy and the energy crisis. Have as many of them share these accounts by reading them aloud as would like to—or have them trade off and read each other's accounts.

Have the group share their reactions to the material from the appendix they have been reading. Has anyone read the article (Appendix B-x) about the problems churches are having in the energy crisis? Do they see any connection between your own church and the churches and situations cited in the article?

—If your church is having an ongoing energy problem, ask someone in the group to write to The Church Building Fund and/or The Church Insurance Company for any data they may have on energy conservation and safety measures in church buildings. (See page 32 for information about these organizations.)

—Whether or not your church has a significant problem with energy consumption, see if it would be possible for your next meeting to have the person or persons in charge of running the furnace and keeping the building warm in general to talk to the group. They will also know, in all likelihood, what the lighting bill is, too. (In warm climates, your church may well have cooling rather than heating problems—energy is involved with solving these problems, too.)

Is there any community energy issue the group is ready to discuss? If they do have one or more, suggest that they try to relate it to any of the Appendix material they have dealt with so far. If there is no issue that comes to mind, suggest that the group bring their fuel bills (of whatever sort) and their electric and gas bills, if they have them, so they can do some comparing at the final session.

Preparing for the Next Session

Start thinking about the last session as a group.

—Is there a local official of city, town, county, state—or even national government you would like to have meet with the group? If there is such a person and they prove to have a busy schedule you could re-schedule the last session, if the group is willing, to conform to the visitor's schedule.

—Just on the basis of what the group has discussed to date, suggest that members make a list for the last session of what they would do nationally to deal with the energy crisis if they were president.

Have the group discuss in a preliminary way the connection between energy and relationships between nations.

—How are nations like people?

—What is the morality of energy? Is there any nation we should not purchase oil from?

—What would you do if you were cold and hungry and could only satisfy these needs by buying fuel and food from a dishonest person at high prices? It is assumed that you could somehow scrape together the money. (It's interesting to remember that these conditions as outlined above were commonplace on the North American frontier in the 19th century.)

—What is an immoral country? Do you believe countries can have morals?

—What is a moral country?

You may want to have the group think over these questions until next time.

Ending

Come together for prayer. Both collect II and collect III on page 259 of the Prayer Book might be used and then the group might recite (or sing) *A Song of Creation* (the Benedicite) which may be found in the Prayer Book in two forms—the traditional form beginning on page 47 or the contemporary form beginning on page 88. It can also be found in the Apocrypha.

And at the Very End

Have the group assess the learnings they have gained in the session.
—What did you learn?
—Identify any new insights you gained.
—What issue discussed in the session would you like to know more about?

SOURCES OF INFORMATION

For information about church facilities and the energy crisis write:

The Church Building Fund
815 Second Avenue
New York, N.Y. 10017

For information about programs to improve the safety and the energy use in your church building or meeting house write:

The Church Insurance Company
800 Second Avenue
New York, N.Y. 10017

Session 4

Bible Reading and Discussion

Begin the session with a Bible reading. The selection from the Gospel of St. Mark which follows deals with the issue of wealth and power. As the group has moved through these sessions, they have gone from their own personal concerns toward, one would hope, a more universal awareness and understanding. As they have moved toward the world view, they have probably become aware of how energy can and does mean or represent wealth and power in the minds of many people and many nations. Someone from the group might like to volunteer to read this selection aloud or the group leader may choose to do it.

As he was starting out on a journey, a stranger ran up, and, kneeling before him, asked, 'Good Master, what must I do to win eternal life?' Jesus said to him, 'Why do you call me good? No one is good except God alone. You know the commandments: "Do not murder; do not commit adultery; do not steal; do not give false evidence; do not defraud; honour your father and mother." ' 'But, Master,' he replied, 'I have kept all these since I was a boy.' Jesus looked straight at him; his heart warmed to him, and he said, 'One thing you lack: go, sell everything you have, and give to the poor, and you will have riches in heaven; and come, follow me.' At these words his face fell and he went away with a heavy heart; for he was a man of great wealth.

Jesus looked round at his disciples and said to them, 'How hard it will be for the wealthy to enter the kingdom of God!' They were amazed that he should say this, but Jesus insisted, 'Children, how hard it is[a] to enter the kingdom of God! It is easier for a camel to pass through the eye of a needle than for a rich man to enter the kingdom of God.' They were more astonished than ever, and said to one another, 'Then who can be saved?' Jesus looked at them and said, 'For men it is impossible, but not for God; everything is possible for God.'

Jesus consistently warns against the danger of wealth. Nowhere does he suggest that wealth or increasing affluence is necessary for salvation or even for the good life. This passage is representative of his teaching about wealth and power.

—Is there a difference between Christianity and the expectations of modern society?

—In a number of the Appendix B articles you have read statements and read statistics that confirm the special role the United States plays in oil consumption. Because of its great wealth and industrial capacity the United States does, in fact, consume most of the world's oil—and more than it produces. Is this right? If you think so, why? If you don't—what do you think could be done about it?

—Must wealth and power corrupt the person or the nation that possesses it?

—You have all read accounts of the desparate need some Americans have for gasoline to keep their cars—and their way of life—going. You have undoubtedly run into the idea that some people will be "willing to pay any price." How do you think a Christian should view this issue? How important a need is it of yours to drive a car? Could you do without one or use one less?

—What is the most important issue in your life that you are willing to share with the group? If the issue doesn't have to do with money or finance of some sort, just how far down the scale do you place those issues? If financial or money issues are your top priority, how do you rank the other issues in your life?

Moving Along

This is the spot to introduce any outside visitor the group has invited to speak to them—whether it be a local person involved in the energy aspects of the church or community, or a governmental or business person prepared to talk about the larger dimensions of the crisis. The group member who invited the visitor should introduce him or her. The visitor may or may not choose to intersperse questions and answers in the discussion. If he or she does not, the leader of the group should invite questions afterward. It would be useful for the questions to be as much to the point as possible, although it is not inconceivable that material might be cited from Appendix B in framing the questions—if it pertains.

If there is no visitor—or even if there is and time is left—this would be a good time for the group members to discuss the lists they prepared for this session about what they would do on the national level to deal with the energy crisis.

—Is there an Appensix B or Appendix A article that seems to offer a serious and important solution to any aspect of the problem?

—Must the incentive for energy legislation and reform come from Washington?

—What issues has the group spotted in their community or state that seem to reflect directly the major national concerns about energy?

—The group has been in session for a good many hours all told. Do they have a sense that there is any impact they as individuals may have on the situation or do they feel that it is still a problem that can only be dealt with at top level?

In the last session, it was suggested that the discussion about morality among nations (topic 6 in the session 3 plan) in light of the energy crisis might be continued into this final session if the group was willing.

—Do you feel a nation can be "evil" or "good"?

—If you came to the conclusion that you lived in an "evil" nation, what could or would you do about it?

—Do you feel the United Nations has handled world crises well? Have they played a useful role in the world energy crisis to date?

—It has been said that the United States would consider going to war to protect the oil supply of the Western nations if an adversary nation threatened to cut it off. Do you think this kind of action would be justified or unjustified? Is there any justification for going to war? What did Jesus say about taking action against evil doing?

—Have someone in the group read the account of Christ and the money-changers aloud (John 2:12–22).

—What is Christ telling us to do when we identify evil in the world?

—But what about turning the other cheek?

This is the last session. Would the group like to share what they have learned? What they have not learned that they want to know?

—Is there any interest in the group for having one more session beyond the contracted period of time? It might be a session to hear a particular speaker who was not otherwise available.

—If there is not to be another session, do any individuals have ongoing plans or projects that have emerged from the sessions? Have they made changes at home? Are they on committees to make changes in the church?

—Does a group member have any "for further reading" advice for the group as a whole? This could either be an Appendix article that was not generally discussed or a book or article or pamphlet available elsewhere.

Ending

There should be closing worship. If the group has expressed an interest in having a Eucharist, fine; but it should not be forced. An alternate possibility would be the beautiful, simple service called "At the Close of Day" from the Prayer Book. It is intended for use by families and it is possible that this group has grown into a family of sorts.

At Close of Day

Psalm 134
Behold now, bless the LORD, all you servants of the LORD,*
   you that stand by night in the house of the LORD.
Lift up your hands in the holy place and bless the LORD;*
   the LORD who made heaven and earth bless you out of Zion.

A Reading
Lord, you are in the midst of us and we are called by your
Name: Do not forsake us, O Lord our God.
Jeremiah 14:9,22

The following may be said
Lord, you now have set your servant free*
   to go in peace as you have promised;
For these eyes of mine have seen the Savior,*
   whom you have prepared for all the world to see:
A Light to enlighten the nations,*
   and the glory of your people Israel.

*Prayers for ourselves and others may follow. It is appropriate that
prayers of thanksgiving for the blessings of the day, and penitence for
our sins, be included.*

The Lord's Prayer

The Collect
Visit this place, O Lord, and drive far from it all snares of the enemy; let your holy angels dwell with us to preserve us in peace; and let your blessing be upon us always; through Jesus Christ our Lord. *Amen.*

The almighty and merciful Lord, Father, Son, and Holy Spirit, bless us and keep us. *Amen.*

And at the Very End
At the conclusion of the service the leader of the group could do all the members a great service by passing out a list of the names, addresses, and telephone numbers of the people in the group so they may stay in touch on the issue or issues they have been discussing—and on other issues if they wish.

Appendix A

Profiles in Energy
These vignettes and the discussion questions which follow them were prepared by a group of interested members of an energy fact-finding group within a major, national church organization. They are not made up. Every one is real. Find yourselves in them.

CITY BOY / *by Oliver B. Garver, Jr.*

I am a total city boy—born and raised in Los Angeles. Eggs and meat come from markets. Water comes out of a tap. Milk comes out of a bottle. Electricity comes out of a wall plug. Temperature is controlled by a thermostat. Gasoline comes from the corner station.

I have read in books and seen movies about chickens and steers and farms and oil wells and coal mines and basins to trap melting snow. I have known intellectually that the delivery systems which assure my survival and satisfactions stretch often great distances back to "primary sources." But—there is "we" and there is "they," and I really don't know very much about them or about what they do, day after day, to feed into my delivery systems and my life. I am simply very thankful that they are out there, doing their thing. And I assume too casually they will always continue to do their thing for me living in my big city.

Part of that hidden outside world was revealed to me in graduate business school. I learned there were lathes and punch presses and giant machinery and assembly lines and workers and managers and profit-and-loss statements. I learned that at the other end of many of my delivery systems there was a huge free-enterprise industrial complex. And behold, it was good!

Indeed, it was an exciting and challenging complex, and I was eager to become a part of it and to rise up in the corporate structures until I could play a major role in the decision-making processes which made it all work—profitably.

After 11 years of this excitement and a good sense of personal contribution and accomplishment, I resigned my position for reasons I can't articulate and entered seminary. My first assignment after seminary was as curate in my home parish—the parish was affluent, comfortable, conservative. I said and did all the controversial things young (though I was almost 40) curates say and do. Then 1964, Selma, Montgomery, McComb. I spent the next seven years of my ministry in a Mexican, Spanish-speaking inner-city parish on the East side of Los Angeles—totally across-town in many ways—in the thick of the Hispanic civil/human rights revolution. "Why," I ask, "can't this better world be more fully theirs *now*, or at least in a more equitable sharing by those who do now have so much?"

Well, who am I today? I am still a city boy, and I always will be. If I had to milk a cow to survive, I'd soon be laid to rest. If I had to mine my own coal to keep warm, I'd be continually cold. If I had to run and maintain my own personal generator for electricity, the TV tube would be a total blank picture. In my city I am an utterly dependent creature for my survival and my pleasure.

There are also hundreds and thousands of poor people of every race and nation living in my city. They and I share this dependency.

We are all dependent on strangers in mines and factories, on farms and ranches, driving trucks and trains, for our daily well-being. Too rarely do we concern ourselves with the daily well-being of these strangers. Or, with those whose lives are significantly affected by the proximity of their homes to our "primary sources" and by the delivery systems passing through where they live, finally to end in Los Angeles. It concerns me that we are exploiting others for our own personal benefit—and we don't really know or seem to care. There is a great rural America out there somewhere which sustains and nourishes all us city folk—but, at what cost to their values and aspirations and well-being?

As Los Angeles and other cities grow and grow and grow, our energy requirements soar. Our utility company sets out to satisfy this increasing demand. Fine. I need electricity in my wall plug. One day, however, I find in my mail an urgent, impassioned plea from the Indian tribes in the Four Corners Area where Arizona, New Mexico, Utah, and Colorado meet. Their land is being exploited and environment ruined to satisfy those power needs. "Whose side are you on, baby?"

Or, plans are drafted for an atomic energy power plant. No city wants the plant anywhere near its borders. "Put it out in the country—far away!" We forget that other people live out in that country, enjoying a lesser density of population and equally fearful of such a plant in *their* backyard. "Whose side are you on, baby?"

Or, present prime agricultural land turns out to be a potential site for a strip mine. What this does to the environment is a well-documented horror story. But, what about the farmers and their lives? Do we push around those who feed us for the benefit of those who warm us? "Whose side are you on, baby?"

I know whose side I'm on. I care about the strangers out there. I want to hear their side of the story. If my dependency results in their being exploited, then I must change my life style. They may be strangers but they are also children of God.

Questions

—On whom does "City Boy" depend for his energy? His food? His entertainment? On whom do you depend for your basic necessities?

—How does being dependent upon others make you feel?

—Does it make a difference in how you feel that you never see all the people upon whom you depend? If so, what is that difference?

—What is the basis upon which you are willing to sacrifice other people you depend on, such as ranchers, uranium miners, coal miners, farmers, in order to maintain present patterns of energy supply?

THEY DECLARED WAR—AND WE ARE THE ENEMY / *by R. Baldwin Lloyd*

It was near midnight April 5, 1977. A college student came over to me and said, "Isn't it awful what's happened to Grundy!"

"What's happened to Grundy?" I asked, aware that Grundy is in the heart of the coalfields of Southwestern Virginia.

"It got washed away—I understand they've really had a bad flood and the whole business area is under water."

It was just a matter of time. We don't seem to learn a thing. What good is it to go to Congress or to state legislatures? Hundreds of people have testified—flooding is bound to happen, they've said, if strip mining continues. It's frustrating and infuriating. I had just returned to Blacksburg, Virginia, from an Energy Ethics Conference in New York, and I was just hearing the first news of what would, in the next few days, prove to be the worst flood disaster in the history of the coalfields.

Once home, I was greeted by my wife, who said, "Orb Caudill called. He said the whole McClure River valley is flooded. St. Stephen's Church of the People in Nora is washed into the river. All the phones and power are out except for the phone at Orb's house. Many people are homeless—don't know the full extent of the damage yet."

Nora is in Dickenson County, Va., adjoining Buchanan County of which Grundy is the county seat. I said, "I can't stand it. Those are my friends. I feel so powerless to help. I know exactly what will happen. There will be lots of sympathy and some immediate relief sent in—and then it'll all be forgotten."

The next morning, news began to come in about the utter devastation of Williamson and other communities of Mingo County, West Virginia. That morning I drove to attend a meeting of the Appalachian Coalition, a project that links together state and local community groups struggling against the severe and adverse effects of strip mining. There were few at the meeting—

because those from West Virginia and Kentucky were flood victims or responding to friends and family who were. Before that day was over, the full impact of the enormity of this flood began to come through. Pikeville, Harlan, Pineville, Hazard, Barboursville, all Kentucky towns; St. Paul, Coeburn, Haysi, Appalachia, Big Stone Gap, St. Charles, Clinchport, all Virginia towns; and many more in Virginia, Kentucky, West Virginia, and a few in Tennessee—over 45 counties hit, and scores of communities.

The gravity of the situation is spelled out in these statistics reported by the American Red Cross: 27 dead (only a miracle prevented the figure from being in the thousands—for most flooding happened while people were still awake); 1,600 small businesses destroyed; 225,000 families suffered major property losses.

First reports came stating 5 to 8 inches of rain had fallen. In a couple of days though, state agency and coal industry spokesmen were insisting it was 10 to 15 inches of rain and that strip mining had nothing to do with it.

"It's an act of God," they said. Can you imagine? Only a miracle of God could *prevent* floods where mountains have been so torn and streams clogged up by the consequences of strip mining. Many areas outside the minefields also suffered floods in 1977. But communities in the coalfields suffered twice the extent of damage that non-coal counties or communities experienced.

The flood was devastating; relief and recovery were slow. In the course of the next nine months, the grueling efforts of recovery in many communities were to be completely undone by successive floods—in some communities as many as three or four more floods. Residents of St. Charles, Va., experienced seven such disasters between April 1977 and January 1978.

Oh how I wished those responsible for strip mining, or who had supported it in Congress, had to live just once with what the people of St. Charles have had to endure. If they did, strip mining would soon stop in most of Appalachia. But they don't experience this, or few do. And they have the audacity to say, "We do a good job of reclamation." And they pass out their shiny brochures with pictures of their showcase jobs.

While most citizens affected feel that strip mining is a major contributing factor, coal operators and their allies and most state and federal politicians flatly deny any connection between strip mining and floods. In spite of early reports from citizens and news media making these claims, and citizens' demands for a fuller investigation of all the causes of flooding—the amount of rain, forestry practices, highway construction, etc., *and* strip mining—most state and federal committees set up to look into the matter sought to whitewash any smudge that might be connected with strip mining.

It is incredible to me that such a reason for investigation should not be responded to immediately from every quarter of government, business, industry and concerned citizenry—and the church. The churches respond well with money and many relief workers, but are mostly silent about the need to get at causes. The need to get at causes seems, yes, eminently reasonable to those who were or might be directly affected by subsequent floods. But to the coal industry people and their allies, and to those who see coal as our paramount need in response to the energy crisis, no, such a search for causes is *not* reasonable. Appalachia's coalfields still produce most of our nation's coal. Our nation's "need for increased energy" and also "for independence from OPEC nations' oil" makes it essential to mine more Appalachian coal as rapidly as possible—twice as much by 1985, the projections say. The quickest way to get this is by strip mining, even now with a federal strip-mining bill.

I can really identify with all those people in the Appalachian coalfields who are getting the feeling, yea the conviction, that there are those in our country who are quite willing to write this region off "for the good of the rest of the nation."

This will give you an idea of the extent to which government and industry are willing to sacrifice a people and their region: While scores of towns and counties and tens of thousands of people were still reeling under the impact of the 1977 April floods, Congressional committees were weakening the federal strip mine bill under the influence of powerful lobbying efforts by industry and coalfield politicians. Can you imagine it! While thousands were still digging out of the mud, assessing their losses, etc., slope regulations which might minimize the degree of erosion or water run-off in areas of steep mountains and frequent, heavy rainfall were thrown out of the bill. And to secure the support of West Virginia's Surface Mining and Reclamation Association for the federal bill, mountain-top stripping, which will compound flood conditions, was legitimized. The bill that was finally passed was considerably diluted from the one President Ford had vetoed earlier and which had failed by 3 votes to gain the necessary two-thirds majority to override that veto. Yet this weakening of the bill happened with a president in the White House who presumably would sign any stripmining legislation sent up to him. It could have been as strong as Congress wanted it to be. But apparently this is what Congress wanted. And so those in Appalachia are right who say, "The region is to be sacrificed to meet the energy needs of our country."

What bothers me the most is how easily now I come to accept these decisions—not that I will ever cease to struggle and hope for better. I know the power of those who own and control our energy resources. And I know how important it is for those corporate entities to expand continually their operations.

I do not believe (as we are told by most in government and industry) that we in the United States need to produce more energy. I especially do not believe we need more energy if the production of that energy is based on non-renewable resources and is going to have such destructive consequences to people and to land, to water and finally to the air. The projected growth in energy use, I believe, is not based on meeting basic human need as such, but rather on the need for and commitment to the principle of "maximization of profit."

We have created two horrendous monsters: one is an industry and an economy that must forever expand production in order to survive; the other is a society that must consume more and more to be satisfied and to insure the survival needs of the first.

I have always been impressed by how the people of the Central Appalachian coalfields will work so hard and sacrifice so much to meet their nation's coal/energy needs. But now they may be less willing than in the past. For there is a growing anger that smolders closer to the surface now, anger because of the lack of provision for enforcement of safety and health measures in the mines. "They ask us to keep cities lit and our industries humming and growing; but they won't assure us of safety and health in the mines or guarantee us retirement benefits, even though we break our health and bodies in a lifetime of mining. We're getting tired of working in danger; and now, with the floods, no place is safe."

One Kentucky man summed it up well for me: "They declared war, and we are the enemy!"

I often get angry—really angry—at God, because these defeating forces continue to crush the lives of so many in Appalachia. Mines continue to be unsafe and unhealthy even with a law (passed in 1969) that, if enforced, could make a difference; more and more people live in an atmosphere of increased dread, in a permanent disaster area, knowing that even with *less* rain, there will be worse and more frequent floods. And even in the midst of prosperity in some areas of our country, poverty for at least half the population of Central Appalachia is made more grinding than ever by inflation, floods and severe winters.

It is clear to me that on the question of energy policies now projected by both government and industry, Appalachia—its people and its land—stand only to lose. Not until we as a people get straight again what our basic relationships are, what our basic human needs are; and what makes for our peace and the peace of our communities, can we expect anything but worsened conditions of poverty, hunger and oppression. Until that happens, we as a people will also become more divided and alienated from qualities of life that really matter.

Our job as the church is to raise the moral and ethical questions. Our job as the church is also to witness to the vision of what it is to be a whole person and to be community. Our job as the church is to show how that covenant relationship between God, God's people, and God's creation gets translated and lived out in our time in history.

Questions

—How do people in Appalachia feel about their region and they themselves being made "victims" so that other people can have more energy?
—Why has the kind of human tragedy illustrated by the events describe here have so little effect upon the legislation and enforcement of strip mining regulation?

—Why are we willing to increase our national (and personal) energy use when it will result in more strip mining?

WHERE NOBODY GIVES PARTIES ANYMORE
by Carolyn Alderson

The plans for the development of the coal reserves in Montana and for the industrialization of this remote rural range country have changed a lot of lives and communities. The changes are not all bad. What has taken place has enlightened people to their responsibilities to each other and to the democratic process. People have discovered new strengths within themselves. People have found out who are their true friends and who are not.

The fear of what is coming and apprehension about it is so great, however, that it is sometimes difficult to see the good parts. This is the story of a journey from a privileged childhood through a period lived as a somewhat ordinary ranchwife and mother, to political activism in a time when divisions in families and communities have become so sharp and final that most of us wish fervently that there had not been jungles and rain forests here sixty or so million years ago—for the coal beneath our land is a legacy from that prehistoric time. Or at least we wonder why the frantic scramble has begun to use up, by the turn of the century, what so long ago was put beneath our ground?

Growing up on a ranch at the base of the Big Horn Mountains was to be the beginning of my closeness to and understanding of land. There were endless opportunities to experience beauty and the interconnectedness of life with all of my senses. I smelled the chokeberry blossoms in the springtime and the new-mown hay in summer; I tasted the wild raspberries and the service berries. I lay for hours in the soft grass looking up at the clouds or hunkered on a stream bank gathering shiny rocks from the creek. I bundled up and made snow forts in the winter. I rode my pony all over the ranch. Solitude was good, easier than dealing with my contemporaries in many ways. I am sure it affected all my later choices of where, with whom and how I would live.

For the first eight years of marriage, I turned my attention away from the tumultuous goings on in the rest of the country during the 1960s. It was surprisingly easy to do living sixty-five miles from town on a gravel road. I had my three daughters, and learned to grow things: vegetables, calves, puppies, and children.

I learned about the real business of ranching from my husband as my father had never thought girls should be in on the nuts-and-bolts (conception, birth and death) processes of ranching. I learned that grass is our principle product, and that the cattle are merely the vehicle to harvest it rather than the primary product. I learned that the success of a ranch in grass country depends more on doing nothing, selectively, than on constant worrying about the cows or the land or the water. This is something most ranchers never learn. The amount of money that goes into large complicated machines to replace people on farms and ranches is mind-boggling. By concentrating on keeping good fences, breeding cattle for disease-resistence and for good instincts for looking after themselves and their young, we have managed to make a little money.

One of the best things about married life was our community. Tiny as it was and still is, no more than sixty of us in a 20-mile radius, there was a remarkable diversity in its members. Many neighbors became close friends. One friend, married to a man thirty years her senior, has given me endless wisdom on child-rearing and people in general. Though she never had any children of her own, she understands them better than most who have them.

Almost everyone in the community is of third or fourth generation. Daughters bring husbands back to the home ranch and sons bring wives. Grandparents are integral parts of family life. They provide refuge from grumpy mothers and they relate the history of the area in vivid personal terms.

The first years after we were married there were big parties, given at the slightest provocation, and everybody always turned out. Again it was a place I felt at home because differences and eccentricities were appreciated. Gentle humor about each other and ourselves abounded. One local "bachelor" claims with a rueful grin that his marriage ended because "she got readin' them Vogue magazines and everyone in them was either sittin' down, layin' down, or leanin' on something."

In 1971 everything changed. Looking back, the change seemed slow. In fact, it was sudden. No sooner had speculators begun coming around asking to buy or lease our coal than the fat government document called the North Central Power Study, was unearthed to us. Many ranchers did lease, thinking it was the same as leasing for oil, which was essentially free money as there was little action after leasing. The North Central Power Study contained blueprints for strip mines, power plants bigger than any presently in existence, water impoundments and pipelines, railroads, and transmission lines. Curiously, it contained no provisions for where and how to house, transport, or in any way care for the thousands of people who would be coming in to do all of these things. It was also terrifying. It produced a strong reaction from people living here, because it was as though those who drafted it didn't acknowledge our existence.

Almost immediately ranchers and other local people from all over Montana came together to form a citizens' group to fight it. We knew little about organizing people, nor did we have any idea where to begin to try to stop the monster. We did know that information, for ourselves and for others, would be our best weapon. We set about gathering as much as we could as quickly as we could. It didn't seem possible that America, which had done without our coal for so long, could suddenly "need" so much of it so fast.

One effect of the development plans on our community was a surprise—and a bitter disappointment. Some of our neighbors were instantly bedazzled by the prospect of millions of dollars they were sure they would have soon. Some naively thought they could have the money and their ranches, too. Others thought that to leave their grandchildren money would mean greater security than to leave them the land that had already sustained several generations quite comfortably.

Splits came about in families, too. One father leased the coal out from under his children. Another threatened to sue his son for standing in the way of a mine proposal the son didn't want. All of this happened as the two men worked side by side in the fields. When I asked one grandmother, "What if your grandchildren want the ranch?" she replied: "My granchildren will have to look out for themselves."

Those of us who were against the development of the coal and who wanted to continue ranching, were often accused of standing in the way of the property rights of others. Or, of being rabid, unyielding "eco-freaks." Out here the term "environmentalist" generally means someone who doesn't have an economic or a lifestyle stake in the outcome of a project.

Our privacy was going rapidly as reporters and sympathizers and company men came round prying into our lives. One reporter asked a neighbor what the biggest change to happen in the community had been since the coal had become a factor. I don't think he really understood the profundity of her simple answer because he didn't use it in what he wrote. She answered, "Nobody gives any parties anymore." I can't forget the last time we tried to communicate with a local family who are on the pro-coal development side of the fence. We had them to dinner and I burned three consecutive packages of frozen vegetables in the debate that ensued. The wife referred to my work with the citizens' group as "your little ecology project" as though it were a hobby or something to keep me occupied now that my children were in school.

My husband, who hates meetings and going to lobby in Helena or in Washington, at first gave me his blessing to go out and fight the good fight. However, before long he began to resent the time I spent on it and would say such things as "You spend more time fighting for this place than out working on it, or enjoying it with me." My children's teacher in school once made the remark that "They must know the housekeeper better than their mother these days." My father thought it was all right for me to fight if there were going to be a strip mine on our own ranch, but thought I should forget about "stopping progress" and "troublemaking" in our state as a whole or in the nation. In a debate at a forum, a coal company person asked me if I had a dishwasher or an air conditioner. I 'fessed up. But I said that if he were giving me the choice of my dishwasher or my ranch, he could come and get it—the dishwasher, that is.

Yes, we, too, use more electricity than we should. In order to take my children to the dentist, buy groceries, or go to a meeting, I have to drive a minimum of thirty miles, usually several hundred. Our house was built in the early 'sixties with no thought to energy conservation, and there isn't a thing we can do about it. We can't move—that's why we fight! We can't shut part of the house off and/or rent the other part; we can't tear it down and build a new one. So we are stuck with being at least a tiny part of our own problem.

There's no turning back. I suppose I don't even have a chance to be a coward again. In the 1960's after college when I got married, I ran as far away from strife as I could, and it followed me here. It's time to dig in and stand our ground. In the course of my efforts I have discovered that there is no place in the world where environmental degradation or abuses—poor stewardship of God's earth—does not diminish human beings in its wake.

Questions

—When we are making decisions which will contribute to changing profoundly our own neighborhood or region, what consideration should we give to the feelings and interests of our neighbors when their feelings and interests are different than our own?

—A governor of Colorado said he would not allow his state to become an "Energy Sacrifice Region" for the rest of the country. Do you believe it is legitimate to sacrifice one region for the benefit of another? For the benefit of the whole?

—Do we feel any differently about sacrificing ranchers in Montana than we do Navajo people in New Mexico? Or people in Appalachia, or Puerto Rico? (See subsequent vignettes.) Are there distinctions or differences here? What is the basis of these distinctions or differences?

I WAS THE ENEMY TO THE PEOPLE
I HAD COME TO SERVE / by Henry L. Bird

In 1973, not long after arriving in New Mexico, my family and I were invited by a Navajo employee of the Utah International's Navajo Mine at Fruitland, New Mexico, for a tour of the strip mining operation. This young man is married and has a family and farms a family plot on the reservation in Upper Fruitland.

We met him and drove out onto the actual mining area. He showed us the cab of a dragline operating crane. The cab was as big as a two-story house. The crane shot into the air several hundred feet and the drag bucket made our VW bus look like a toy roller skate. The dragline was used to scrape the topsoil off the fragile desert land to expose the coal.

The coal would be scraped out of the exposed seam by bulldozer, loaded onto huge trucks (and now by a special railroad, too) and taken to a crushing and conveyor-belt operation. From there, the processed coal would be laid out in long mounds the size of football fields, and later conveyed as ordered to the Four Corners Power Plant adjacent to the storage area.

As we drove back out of the mining area, our guide pointed out a dead tree trunk and the ruins of a Navajo hogan about a hundred yards off the road and completely overshadowed by huge piles of dragged-up topsoil. Quietly he told us that in his boyhood his family sheep camp, the place where he had grown up, had been there.

A little farther on we came to a Navajo family settlement with hogan, homes, sheep corral,—and in the distance, juxtaposed against the traditional scenes, the Four Corners Power Plant. I slowed down and said I wanted to take a picture. The scene caught me in the gut. It so clearly showed what was happening to a whole people and their way of life. This unbelievably massive technological structure spewing forth sky-covering clouds of smoke seemed to be poised to devour, as its sacrifice, this Navajo family settlement. Our guide seemed uncomfortable at my suggestion and asked me to go on about a mile. "You'll get a better picture up there." I sensed he just didn't want me to record that scene. So I didn't.

Another day in the spring of 1974 I was asked by some friends of the Coalition of Navajo Liberation to go with them to a chapter meeting at the Huerfano Chapter House. (A chapter is a geographical political subdivision of the reservation.) The meeting was concerned with explaining to the people living in the Huerfano area why some of them would have to move to make room for the Irrigation Project. When we arrived the room was filled with Navajo families.

In the front of the room, on the raised platform, much like one would find in a junior high school, were three men, two Anglos and one Navajo. On easels behind the men colorful maps were displayed. The two white men introduced themselves as agricultural engineers working for the Irrigation Project. They would do the explaining and the Navajo man with them, also an agricultural engineer, would translate.

The men described, using the maps and a series of slides shown on a screen, how the agricultural project was developed and what it would do, the crops it would grow, and so on. The presentation included a description of the Navajo dam and lake some 50 or so miles to the northeast and the aqueduct system that provided the water for the Irrigation Project. After this presentation the men then eased into explaining that this would mean that some of the families in the area would have to be moved away. However, these families need not worry. They would be compensated financially and the compensation would be based on the number of sheep and goats they owned—so many dollars per sheep and goat unit.

Then one of the Anglos said: "You're all 'free, white, and 21'—so you can do whatever you want with your money." It was like a thunderbolt among us, but he didn't even pause and was obviously totally unaware of what he had said.

He then went onto say that one family they had moved had so many sheep and goat units that they received something like $50,000 and that instead of giving them the money outright, they set it up in a trust fund. (Had he said "free"?)

I began to feel much like a person watching the Nazi invaders of France in World War II explain to a French village why the villagers would have to leave their homes—and have a local Frenchman translate for the villagers. The discomfort permeated the room—except for what seemed to be the total obliviousness of the Anglo presenters to the feelings present in that room.

The people were given an opportunity to speak after the lengthy presentation. A woman came forward and with anger in her voice spoke strongly in Navajo and gestured at the maps and in the direction of the dam and lake. My Coalition friends translated for me: "That dam and that lake are not 'Navajo' Dam and 'Navajo' Lake; they are a place for the white man to play around and go fishing in his boats! This is my home; it is my grandmother's and grandfather's home; it is my family's home. I don't want your money!"

Following her, many others stood up and spoke in like manner. Finally, one of the Coalition members asked permission to speak and tried to sum up in both English and Navajo what the people were feeling and how this project really didn't seem to take the people into consideration. The Navajo Irrigation Project staff person, uncomfortable it seemed, tried to express the way in which he felt the project could be helpful to the people. But this didn't ease the tension or frustration.

It had been a big change for me and my family to move from the New England seacoast to the desert-canyon-mountain country of northwest New Mexico to begin work with the Navajo Episcopal Congregations in January of 1973. We have often said since then that it has been like "going to the moon." I had been raised in a segregated city, Wilmington, Delaware, during the Depression. Although we were not wealthy like many of our friends and my mother had been forced to go to work, we still were clothed, fed, housed, and educated in a privileged society and were shielded from the harsh realities that we could sense were out there for others. Later, in boarding school and college, I was increasingly aware of social barriers until it seemed that wherever we are, we find efforts to pit human beings against one another.

After ordination there was suburban parish, later the Episcopal parish on Martha's Vineyard, Cape Cod, amid many of the elite of American society and also among Gay Head Indians, people of Portuguese descent, a year-round working population, retirees, summer home-owners, summer tourists. A prolonged strike of ferry employees sharply sensitized the population to its dependency on modern conveniences and to the gap between master and servant, powerful and weak, rich and poor. Still later a ministry on the Maine seacoast where some of our high school students were Passamquoddy Indians and later as director of the University of Maine's UPWARD BOUND program I became familiar with the Penobscot people. All of these experiences seemed to open the door to our involvement with the Navajo. And all of them seemed to be suggesting that our gluttony for the gifts of the earth was the power that led us to separating ourselves from each other.

Among the Navajo we found two societies rubbing constantly against each other, one feeling it was superior and had the right to rule the other. And the other, though physically defeated, refused to give in. Surface contacts were always somewhat shy and smilingly polite and respectful. But occasionally, when I would pick up someone who had been drinking, the venom would pour out. For the first time, I was the enemy to the people with whom I had come to serve.

I represented to them the power that had taken control of their land, that hungered for the mineral resources—coal, uranium, oil, and gas—and would do anything to get them. They could see and feel their traditional way of life deteriorating in front of them. There is pain when older people begin to realize that the children do not fully understand their language. There is pain when a traditional ceremony is attended by a drunken, disrespectful youth. There is pain when a teen-age girl is found crawling at a fair ground in the early morning after an explosive device forced into her by a gang of young toughs had gone off in her vagina. There is pain when 40 to 65 percent of the population cannot find work. There is pain when three indigent Navajo men are murdered by three white high school students and the community erupts with five successive Saturdays of demonstrations, with threats of shootings and violence. There is pain when a young man comes to you with a rifle and belt full of ammunition just before he loses control and in an expression of hurt and frustration shoots anyone he can find.

It is at times of such pain that I hear ringing in my ears the Eucharistic words: "This is my body . . . broken for you." The Broken Bread and Poured Cup speak so deeply of the broken and poured-out humanity in the world around us, where the Lord is and where we too are called to be in His Name. "Give us this day our daily bread" we pray. Yet the bread of life is broken in the getting and in the eating. And the wine of life is as often drunk in pain or sorrow as in joy. Yet Our Lord said: "Drink this, all of you." And he told us "This is my blood of the new covenant . . . shed for you and for many." "Do this," he told us "in remembrance of me."

It is not all pain, although there is pain when white people say: "Why are you with them against us—your 'own' people?" But there are also touching moments that cross barriers. A groom gives you the blanket he wore at his wedding; a woman gives you a shawl for your wife that she has made; a teen-age boy who had been seriously stabbed by a drunken brother and whom you had seen in jail and in the hospital comes to you at Christmas time, saying very little, but putting around your neck a silver and turquoise cross on a woven leather cord, both of which he had made himself.

Energy crisis? Is all this coal the answer? Does the huge Navajo Indian Irrigation Project and Navajo Agricultural Projects Industries really serve the needs of the people—or of agribusiness? People are told to move off their land and sheep camps to make room for it and they don't understand, but they have no choice. Water is needed for population growth, for agriculture, for power plants, for mining. A whole city of 4,500 is being built and will be populated within a few short years where a few scattered sheep camps used to be. Does this mean a new life for these people, or does it mean more confusion, uncertainty, upheaval?

Fifty percent of the Navajo population is under 18. What is their future? Are they to be Navajo, true to themselves and their heritage? Or is that no longer possible? What about the anguish and anger of the young Navajo who cannot speak their own language?

It is not just a question of energy resources here. It is a question of life and how we share it—or do we only seek it for ourselves?

Questions

—The costs of greater energy production, such as the sacrifice of Navajo land and culture, have been justified in terms of achieving "the greatest good for the greatest number." Do you agree with this way of thinking? How would *you* judge the various goods involved, and how would you calculate the relative balance?

—The costs of greater energy production, such as the sacrifice of Navajo land and culture, are sometimes justified in terms of the cash compensation the Navajo people receive for losses incurred. Can you imagine how much compensation you would feel appropriate if you were losing your society and culture and ancestral lands?

—The sacrifice of Navajo land and culture for the sake of greater energy production might be defended on the grounds that the Navajo had a choice and took the compensation. What factors do you think probably entered into their decision? How much of an option do you imagine they had not to settle for the compensation?

GUÁNICA BAY AND THE INVASION OF PUERTO RICO
by Benjamin Ortiz-Belaval

I was born here in Puerto Rico some forty-four years after North American troops invaded our island on July 25, 1898. The troops landed precisely through the Guánica Bay—to the southwest of our Caribbean coast. Guánica Bay today is the "capital of pollution" in the Caribbean, for companies like the Commonwealth Oil Refinery (CORCO); Pittsburgh Plate Glass (PPG); Union Carbide; Rico Chemicals; Puerto Rico Olefins and others have settled precisely in that original invaded neighborhood. So some of us in Puerto Rico see the energy problem as but one more episode in an outgoing and imbalanced struggle over the defense of Puerto Rico and its natural resources.

Since that first invasion in 1898, whatever development has taken place in Puerto Rico's economy has been oriented toward satisfying the market needs of the U.S. corporate structure. First it was sugar. All of our island was cropped for sugar. Even our mountains. The first imports of oil were to provide energy for the sugar mills. This first economic stage of development ended in the late 1940's.

The second stage was light industry. Puerto Rico's work force of cheap and unskilled laborers provided an investment paradise for the garment and shoe industries. The local government paved the way by (1) taking over the local energy company (Autoridad de las Fuentes Fluviales) and by (2) granting total tax exemption to whoever was interested in establishing an industrial firm in Puerto Rico. This was the period of Operation Bootstrap, marked by a cheap labor force, a very low-cost production of light goods, and an unexpected invasion of foreign capital (though this time from South and Central as well as

North America), capital which would circulate in our economy but not accumulate there. This period ended in the late 1960's.

In this second period of light industry, oil was again needed by the Puerto Rico Water Resources Authority in order to provide electrical energy to those firms. The first big polluter of the air and waterways of Puerto Rico had been the sugar mills, but now the energy authority, Fuentes Fluviales, became the island's greatest polluter.

Parallel to the establishment of light industry there was the birth of a local consumer-market for U.S. goods ranging from GE microwave ovens to Del Monte canned green beans to air conditioners and Ford Pintos. Right now Puerto Rico is the fifth largest market for U.S. goods in the world, and the second largest market in Latin America.

Then came the last and worst stage, the establishment of semi-heavy industry in the form of highly technical chemical, petrochemical, and pharmaceutical enterprises. These were the so-called "blue chips" of the U.S. corporate structure, and they were capital-intensive, highly polluting and produced few new jobs. These companies also have tax exemption. They have a labor market three times cheaper than in the United States. And they have presently a new incentive—the government advances them 25 percent of the labor cost and provides all of the infrastructure systems for the new facilities. That means free water from our watertable or groundwater system, cheap electricity (i.e., cheap oil and special electric rates from the electric utilities), and a beautiful sky to look at and to dump wastes into.

Of course there are no such firms in the tourist section of El Condado-Isla Verda. To pollute you locate always in poor "arrabales," neighborhoods such as Guánica Bay.

Four huge refineries were established to support this capital-intensive, energy-intensive, and profit-producing paradise. More than half a million barrels of oil from Venezuela are processed every day by the refineries of CORCO, Phillips, SUNOCO, and Gulf, either to be used locally or to be sent on to the mainland of the United States by way of the Gulf States.

During this latest period a complete inventory of Puerto Rico's natural resources was undertaken. As school children we had all been taught that Puerto Rico was very poor, that it had no gold or silver, no copper or nickel, and especially our island had no oil. The message was that, in order to survive, you had to learn English, to go to New York, and to depend on their goodies to survive. They would take care of your nature, both inside you and around you.

Suddenly in this latest period those school children learned the results of the mineral exploration of Puerto Rico which had just been completed. There were big deposits of copper in what was now being called the "mining triangle" of Utuado-Lares-Adjuntas. There were big deposits of nickel in the west coast town of Cabo Rojo. Big deposits of gold in the central mountain town of Corozal. And since 1973 big deposits of oil have been found to the north of the coast of Manati, Dorado, and Old San Juan itself.

But at the same time we learned that we could not deal with these resources by ourselves, and that we would need the "support" of American Metal Climax and Kennecott to mine our copper, the support of Universal Oil to explore and develop our nickel, and that we had to give exploration permits to St. Joe Minerals to take over our gold. And who better than Mobil or Exxon or Superior to "manage" oil exploration and exploitation for us? So you were poor and now you are rich! But let us help you become really rich by exporting to you our technology, our know-how.

I am the Executive Director of the Puerto Rico Industrial Mission (PRIM). We tried to stop AMAX and Kennecott from making the "steal of the century" in copper. Up to now (September 1978) we have succeeded. Then it was an oil superport they wanted to plunge into our west coast. We stopped that, too. Then it was a nuclear plant—"If you do not like oil processing because it pollutes, then why not nuclear energy?" For reasons we nowadays do not even have to insist upon (but at the start we had to help people learn about), we opposed nuclear power plants and, to this moment, have stopped them. Then came the mercury intoxication by a thermometer plant in Juncos, again far from all the tourists at El Condado-Isla Verde. The company finally paid $2.5 million to the neighbors.

Today the struggle and the invasion of Puerto Rico continue. The pharmaceutical complex to the north. Rum and oil "distilleries" in the metropolitan area of Cataño. Nickel exploration and tuna fish wastes to the west coast. Oil refineries to the southwest. Copper in the central-west mountains. Gold in the eastern-central mountains. Another oil refinery and chemical-pharmaceutical complex to the southeast. Union Carbide Graphite to the east. And the U.S. Navy bombarding and destroying Vieques, an island-township off the east coast of Puerto Rico.

It all stinks. And there is almost nothing left for us.

Questions

—What do you think of the claim that Puerto Rico has been "invaded"?
—How do the perceptions in this vignette compare with those of the energy corporation executive?
—Do you agree with the understanding of progress assumed in this vignette?

THE WELFARE OF HUMAN BEINGS . . . DEPENDS ON THE MAXIMUM USE OF NUCLEAR POWER / by Stanley E. Turner

More years ago than I now like to think about I walked through the doors of one of the world's largest oil companies bearing a fresh Ph.D. diploma and embarked on what I thought then would be a life-long career in finding and developing new sources of oil. I was to begin about that same time what was to be a life-long parallel career in the Episcopal Church, first as a lay person and lay leader in local and then diocesan settings, now as one preparing for ordination soon to the diaconate.

After five years with the oil company I had been impressed with the validity of the company's claim of a coming energy shortage, with the projections of excessive oil consumption, and with the fallacy of our nation then moving to become dependent upon foreign sources of oil. I also recognized the importance of a new and then undeveloped energy resource, nuclear power, and in 1957 I made a decision to shift the focus of my work from oil to nuclear energy.

During the early years most of the effort in nuclear power development was devoted to evolving concepts that would be economically competitive with other energy sources. And with developing an understanding of a myriad of safety concerns—and resolving them. Today nuclear power offers the cleanest, safest, and most economical source of power yet available.

My own involvement over the past twenty years has been principally in the area of nuclear safety and reactor core performance—first, as a member of a design engineering company and now as an officer and founder of an independent energy consulting firm. The safety standards that the nuclear power industry strove to develop far exceed those used in any other industry. In spite of the extreme care taken and the outstanding safety record, voices have been heard in opposition to nuclear power raising safety concerns of various kinds. Some of these were valid concerns. They have resulted in improved safety, for as in any industry, improvements are always possible. Others of these concerns were only misunderstandings, exaggerations or issues considered out of context. And occasionally there have been deliberate distortions and untruths.

All facets of our lives involve risk, sometimes recognized and sometimes not. Perhaps one of the greatest fallacies in reasoning today is to single out one aspect of our lives—such as nuclear power—and attempt to evaluate its risk on an absolute basis, either without regard for the relative risk of alternatives, or to look at those alternatives using completely different criteria. For example, it is a well-documented fact that a nuclear plant emits less radioactivity than a coal-power electical generating plant of comparable size. And such a comparison says nothing about the release of poisonous fumes and smoke when coal is burned. Nor does it take into account the so-called "Greenhouse" effect, namely, the long-term alteration of the climate by the carbon dioxide released in burning fossil fuels. So what I am suggesting is that any valid assessment of the risks of nuclear power must necessarily also assess the relative risks of the alternatives too, and that the criteria for evaluation should be comparable.

Much has been said about nuclear wastes. Actually, the nuclear wastes are many, many times smaller than the wastes from a coal plant—a few hundred pounds compared to thousands of tons. And radioactive nuclear wastes are less hazardous than many industrial materials now in common use throughout the country. So again, by considering the issue out of context, it has been possible to create the impression that nuclear wastes are far more hazardous than they actually are. Disposal of nuclear wastes, as demonstrated in numerous tests and pilot plants, is not a serious problem.

To the credit of the nuclear power industry, it must be said that much more is known about nuclear power plant operation and the consequences of abnormal circumstances than for any alternative energy source. The few comparative evaluations that have been made clearly show that fossil fuel plants are more hazardous and have a much greater adverse effect on the environment.

Opponents of nuclear power often become selectively preoccupied with emissions from nuclear reactors which are almost negligible, and magnify them far out of proportion when they are compared to the natural radiation which exists or the relative consequences of other private or industrial operations of comparable or greater hazard. The same is true of postulated accidental releases of radioactive materials from nuclear reactors. To see how different the criteria used are for operations of comparable hazard, consider that in the transporting of spent reactor fuel, the shipping cask must be able safely to withstand a broadside impact of a locomotive traveling at 60 mph with no resulting release of radioactivity. However, tank cars on the same train carrying extremely hazardous materials (such as chlorine) are not subjected to those same rigorous requirements, despite the dire consequences of some recent derailments involving chlorine gas. It should be patently obvious that any comparison of the risks of nuclear power with alternatives must be made on comparable bases if it is to have any credibility.

Questions of the morals and ethics of nuclear power have also been raised. Dr. Margaret Maxey (Professor of Ethics at University of Detroit) has pointed out that the use of one-sided condemnatory bias and deceptive and exploitative tactics is ethically indefensible and irresponsible. This is especially the case when used as a political weapon to promote ideological presuppositions without consideration of the welfare of human beings and in violation of social justice. I refer specifically to Shinn and Sigwalt's writing on "Social Ethics of Nuclear Power" in the World Council of Churches publication *Facing Up to Nuclear Power*. The National Council of Churches several years ago received a report on the "Plutonium Economy" which opposed it, without regard to alternatives.

If care is not taken, ethical and moral issues will become dictated by political ideology. To what extent can this country and other affluent nations morally deprive less fortunate nations of the world of needed energy? This is an ethical issue that must eventually be addressed. Some people may feel that they and others are victims of high technology or of an energy-intensive economic structure—rather than its beneficiaries. Such a position may reflect little more than a severe case of historical amnesia, and must be questioned.

In conclusion I want to say that I cannot in any way agree with those who seek to obstruct nuclear power or impose their political ideologies "based on a fanatical belief that this will force a vast economic and political system about a redistribution of wealth and power" (Dr. Maxey). The likelihood of achieving those goals is small and unpredictable. But the social impact and the tyrannical power necessary to bring about such a revolution are truly awesome. As far as I am concerned, the welfare of human beings on this fragile planet and of unborn future generations, depends upon the maximum use of nuclear power now and a continued search for new energy sources.

Questions
—How do you feel about nuclear power?
—Why do you suppose nuclear power has such passionate proponents and such passionate opponents?
—The author of this vignette is a nuclear expert. How are you and other lay people to make up your mind about energy policy when the experts disagree?
—The author suspects the supporters of nuclear power are motivated by "ethics" and the critics by "political ideology"? What do you suppose in the author's mind distinguishes an "ethical" position from a "political ideology"? Do you agree with that distinction?

KANSAS AS RADIOACTIVE GARBAGE DUMP FOR THE NATION? / *by Douglas Mould*

As a young boy in the mid-nineteen-thirties, I remember the very strange, mysterious device that stood upon a hill close to Ardmore Boulevard, between Wilkensburg and Turtle Creek, near where I lived in Pennsylvania. I had asked my father about the large teardrop "thing." He told me it was an atom smasher the Westinghouse Company had constructed. I think that was the extent of his knowledge at that time. For in those days there were few people, at least in our circle, who knew much about atomic energy.

I learned later—in high school—about atoms and molecules, and about the cyclatron Westinghouse had built. My teacher was quite skeptical of the effort, for after all, he said, the universe was neatly ordered with its atoms and molecules, so why interfere with the way of things? Never do I remember him talking about the benefits of splitting the atom. Nor did he convey to us the potential dangers of nuclear explosion and radiation. I wonder now if he either knew or even suspected.

My next recollection of the atom is when I was a fighter pilot preparing to ship out to the Far East. One morning in August 1945 *Stars and Stripes* announced to us the dropping of the first atomic bomb. We were elated. We thought that rather than having to invade the Japanese home islands, we might now go home. The horror of the destruction was made known a few days later. Few of us really cared. They had started it, we felt, and we had finished it!

I became an Air Route Traffic Controller after the war. I was on duty the night after the Korean conflict began. Nineteen B-50s had landed at Kirtland

AFB earlier that day. We received flight plans for them to fly from Kirtland in New Mexico direct to Goose Bay, Labrador.

As the planes were so heavily loaded it was necessary to warm up the engines, taxi to the end of the runway, and then shut down the engines to "top off" the tanks before takeoff. Some of the B-50s got halfway down the runway, realized they couldn't make it, reversed props, and started all over again. When they did get off, I'm sure the planes were only a matter of inches off the ground as they crossed the end of the runway. They settled down into the Rio Grande valley and were lost to sight. But eventually we could see them, like overstuffed birds, staggering into a climb.

We all knew each craft carried a nuclear device—unarmed, of course. But we were concerned only about the people in the valley and about the crew, if one of the planes were not to make it into the air. I felt no anxiety about possible release of radioactive material in a possible crash.

My father by this time was head of an inspection department at Sandia Base and knew more about atomic energy than I'll ever know. He had learned a great deal since my youthful questions about the teardrop structure above Ardmore Boulevard. People who worked at Sandia and particularly Los Alamos were people of prestige and to be respected. They were ushering in a new age of the atom. They were the new messiahs.

It was not until I moved to Kansas in 1965 that I got my first negative vibes about atomic energy. There are salt mines in Hutchinson and Lyons, Kansas. The Lyons mine, no longer being worked, seemed, to AEC researchers, a natural repository for radioactive nuclear wastes.

There were those in power who wanted the benefits of a new enterprise. And there were also those who said, "We are not so sure. We do not want Kansas to become the radioactive garbage dump of the nation." The original AEC work at Lyons had been a research project to study feasibility, and it gained quite widespread community support as research.

When suddenly it was proposed that radioactive wastes be actually stored in Lyons' salt beds, we learned that those who didn't want Kansas to become the nation's dump for radioactive wastes had a point. We also learned that the storage canisters used for radioactive wastes at other sites hadn't proved to be leakproof as had been originally thought. The radioactive wastes would stay radioactive for many thousands of years, and there was disagreement among Government authorities as to whether the wastes should be irretrievable—so no future generation would ever be able to release the waste by accident—or whether the wastes should be accessible enough to retrieve if something unforeseen went wrong, or if one day a use was found for radiactivity.

Today we in Kansas who live near the salt mines at Lyons and Hutchinson find ourselves in a dilemma. The experts don't agree, and, furthermore, we have discovered that those experts whom we had thought were stating facts we later found were stating opinions or "considered judgments." So now we question all those who are very sure they know. And when we think about these things, we aren't sure.

Questions

—To what extent has the passage of time clarified the risks and costs of various energy alternatives?

—Does this "time lag" of perception about problems associated with progress suggest better ways of making decisions affecting the future?

—How do we justify protecting ourselves and our own neighborhood from hazards of energy production and use, while we accept and use energy whose production has subjected others to hazards?

WHICH WAY TO A SOCIALLY-CONCERNED AND USEFUL LIFE? / *by George McGonigle*

I have spent the past twenty-seven years as an employee of Exxon. During none of this time have I been directly involved in the energy-related operating functions of my company. Nevertheless I have been in an executive position for the past ten years or more, so that general corporate concerns, while not my direct responsibility, were still within my daily realm of consciousness.

I lived through the era of gasoline and tire rationing during World War II. All of us felt then that if we could only get enough gas to travel, how wonderful that would be. Next came the natural gas pipelines from Texas to the Northeast that put the coal industry in decline. Then came the Interstate Highway System with its boon to high-speed auto travel. Then jet aircraft—whooshing us to the far corners of the world on wings fueled by kerosene. Then the wonders of nuclear-powered electricity power plants and warships.

Al during this time I felt I was part of the miracle of America and a system of free competitive enterprise that were models of justice, efficiency, and the ingredients of a good life. My church involvement, I feel, supported me in this belief.

But now the validity of my experience is being called into question by strident voices within the church and in society at large. Was this experience, as I had thought, one based upon wholeness and justice? The implication is that my experience was not, and I am to be assigned personal guilt for my failure to perceive the violence my lifestyle was wreaking on the natural environment. Similarly, I am to be assigned personal guilt for the injustice and oppression my lifestyle was inflicting upon the poor of the world.

I am having a hard time not only with the substance of these challenges but also with my assumption that within the church these challenges will be made in loving concern. It seems I am being asked to deny my own humanity, my place in the sun, so that someone else can have not a place *like* mine but *my very place!* It is as though "quality of life" were being viewed as a fixed quantity, and those who are "Have's" must give it up so that those who are "Have-Not's" can get their share.

So it is difficult for me to isolate "energy issues." I see myself day-by-day living within a community of Christians and other friends and opponents. It seems to be a panoply of forces I pass through daily in my quest to live a responsible, fulfilling life liberally salted with seriousness, frustration and fun.

My sense of myself is that, like most of us, I have tended to accept as part of the structure of reality a very large area of life as "given" or "taken-for-granted." What I have concentrated most of my attention and work upon has been what seemed like the margins of what is given, where uncertainty and change presented major challenges. In doing this, first as an engineer and later as a corporate manager, I have seen how the intentions and efforts of many socially concerned Americans have been directed within a large multinational energy corporation toward turning latent natural resources into dynamic, usable power.

Making natural resources usable has seemed to many of us an important contribution, because this energy has been needed to create wealth, and wealth in turn generates the potential for empowering the powerless. What I—and many like me—have believed is that "quality of life" is not a fixed quantity. Instead, by using energy in combination with human skills and organizations and creativity, it is possible to increase the size of the pie. We have seen ourselves helping the cause of justice and the poor by making more of what there is to share.

Yet as I look at the broader scene, I hear fellow Christians who are as dedicated as I and as well-intended as I saying that not only my company but the whole political and economic system of which it is a part are hopelessly corrupt and corrupting, that there is still injustice, and that we are part of the problem rather than part of the solution. These are hard judgments for me to hear, and I wonder about this.

I ask myself, Where in this confrontation is the church? Isn't the church supposed to be the reconciler, the patient loving healer of wounds? All too often it appears to me that the church is on the other side from me, rather than standing in the midst of this confrontation with arms reaching out to bring us to mutual understanding and reconciliation. Do I see this incorrectly? Or is my vision 20/20?

I guess what I'm trying to say is that I feel all of us, whether energy companies, environmental activists, or social-concern professionals, encounter the questions of distritbution of wealth and power, and all of us tend to become instant experts. It seems to me that all of us can learn from the life of Jesus, a carpenter's son, a healthy skepticism for all neat solutions to problems such as justice, poverty, race, economics, government.

What I am coming to believe is that the structures of life ("principalities and powers," if you will) are too subtle and interwoven with the values of truth, beauty and wholeness for us to separate them out and seal ourselves off from them while we try to maximize abstract virtues such as justice or equality which we also cherish. It's like jumping into a pile of dry leaves while holding a wet lollipop and hoping to walk away with no leaves sticking to the sucker. You can't do it!

So what are we to do? I have to rely on my own faith that as we go one step at a time, the ends we seek to achieve are going to grow out of the means we use to achieve them. To me, this means that beginning where one is *at* to do what one *can* may be more significant than having a comprehensive strategy of world salvation. Jesus never told his disciples to reorder their society or its economy. Oversimplified, he told them to love one another and to go forth and make other disciples for the Holy Spirit to lead into truth. So I walk in faith that in small and incremental ways my choices will be enlightened by the Truth, and that I shall be guided toward that perfect freedom that comes from serving God through serving others, by being within the fellowship of the followers of the Crucified.

Questions

—How does this corporate executive see himself as a victim?
—Should the church always be a reconciler and avoid taking sides in disputes?
—How do you think the energy executive would respond to the human costs of energy production and use described in some of the other personal vignettes?

THE CONTRADICTIONS IN SELF
AND ENERGY POLICIES / *by Martin M. Glesk*

Most of my thirty-six years seem to have been spent living up to the expectations of others. The fulfillment of these expectations has left me strangely unsatisfied and undernourished. I find that current attitudes towards energy seem a striking parallel to the course of my life: going along with the system, living up to the expectations of others, not raising the questions or addressing the doubts. I note with concern that the same type of inconsistencies and contradictions which have pervaded my life I find also in our energy-intensive society.

I was raised in a small town in the mountains of northwestern Pennsylvania, growing up in the 1950s, educated as an engineer at a southern university, into the Navy and Vietnam, marriage, Harvard Business School, two children, an interesting, well-paying job, a large home in the suburbs. My work these last seven years has been in energy, the last four in energy conservation and solar energy. It is not that I have the energy solution or that it is my mission to impart some solution to others. Rather, I am conscious of how much I continue to be a part of the energy problem. And I'm aware how much my personal witness in energy matters is undermined by my personal inconsistencies in the way I continue an energy-intensive pattern of work and life.

I said earlier that most of my life seems to have been spent living up to the expectations of others—parents, school, church, college, U.S. Navy, corporations, society. Behind this pattern of striving for success in others' terms I've always felt both a sense that my life seems normal *and* that there is something out of order. Going to engineering school in the South and then military service in Vietnam did not encourage raising questions which seemed dangerous distractions. After the service, going to Harvard Business School again provided me with a pattern of aggressive, success-oriented, intelligent work in which success was clearly measured in narrow terms of money and power.

Like my other decisions, marriage to a girl from back home was the logical thing to do; our families were friends and we had known each other for years. But unlike my other decisions, this one proved to be a decision which eventually changed the direction of my life.

Where I had accepted, she questioned and challenged. It began to be important to me and to my conscience to recognize the implications of my decisions and actions. Our children too helped to open me up further. The self-centered, independent, success-oriented, over-committed direction of my life began to seem more and more out of kilter as it became increasingly clear that the price of my "success" was the oppression of my wife and the neglect of my children.

My new work, consulting, was intellectually stimulating and I became involved in a broad range of energy-project evaluations: strip mining of coal, liquified natural gas imports, oil and gas explorations, electric power load forecasting, energy supply and demand analysis. My energy experience was to lead after several years to evaluations of the potential for energy conservation and for solar energy.

Looking back now, I realize that my early consulting work in the energy field was based upon the unexamined assumption that the survivial of our growth-based economy required the consumption of more and more energy year by year. But my research strongly suggested that growth in energy consumption in the U.S. was not, as we say, "causally linked" to growth in GNP. My policy work for the government gradually disclosed to me other tilts in our policy preferences with regard to energy.

I became aware, for example, that a homeowner's investment in insulation might save so many barrels of oil a year, but it was not regarded as being as desirable or significant as a massive corporate investment in offshore oil or gas exploration to increase energy supplies by the same barrels of oil a year that could be saved by thousands of homes being better insulated. An increase in energy supplies—by massive corporate investment in new generating facilities, or coal mines, or other major projects to increase energy supplies—was in policy terms preferable to reducing energy use and demand. Of course there were the standard arguments that individual homeowners were financially unsophisticated (preferring saving accounts at 6% interest to insulation with 30–50% return). Homeowners were also seen as technically unsophisticated, unreliable, and there were thousands of them who would have to install insulation to equal the effect of one new power plant or new oil well.

But I came to see that such arguments were compelling only if you accepted them at face value. Yes, corporations are more capable than individual homeowners of undertaking sophisticated technical and financial evaluations. But I came to see that the peferred energy policies were not so much sensible as the ones we could do more easily. They were policies adapted to the fact that our capital markets were better adapted currently to supporting a few massive investments than they were millions of individual investments in home improvements and insulation. Our preferrred policies reflected also our feelings about "growth": "growth" seemed good to people and implied progress and prosperity, while there was an appearance of stagnation and a lower standard of living associated with using less energy and reducing energy demand by becoming more energy-efficient.

So the orthodox or standard way of viewing the energy situation was to see it as a problem of getting enough supply, a problem best addressed with technical solutions. This view is very compelling and very pervasive, and it appears to be deeply rooted in an American faith in technology and faith in how "the system" worked.

My own personal life was leading me to an increasingly strong skepticism. I now was regularly anticipating that the apparent "standard" solutions might *not* be solutions but might be obscuring real solutions. One associate, an optimist, believed the standard solutions: problems with nuclear power would be resolved, ne energy technologies will evolve, and our system as we now have it will change slowly and without pain or discomfort or discontinuities. On the other hand, I find myself wondering if we might better pursue the unorthodox way of becoming much more energy-efficient and less energy-wasteful. I ask myself what if my associate's optimism proves not to have been justified? What will we have foregone but the presumed pleasures of high energy consumption, if we begin to act *now* as though our resources *are* finite, and if we do *not* apply technologies until their safety, environmental, technical or societal problems are resolved?

I am convinced that there is substantial risk in the orthodox energy solutions of run-out, melt-down, blow-up, brown-out. So I see that, if my children are not to be left with a wolrd depleted of energy, or dependent on large-scale, highly-centralized and dangerous technologies (whether nuclear power plants or solar power satellites), then radical change from our present energy patterns is essential.

But how consistent is my present life with these energy beliefs of mine? I find my own work and my lifestyle seem unfortunately rife with inconsistencies. On the one hand I feel that in order to change the system, I must labor in it continuing a personal status quo with many problems for me. On the other hand I believe that personal consistency and thoroughgoing integrity of beliefs and actions may be possible only through a radical alteration in my work and my life.

My wife and I have had a number of conversations about this, especially after I had worked nights and weekends for several months of intensive effort on a solar project for the Department of Energy. She said to me:

"I see that Carter has cut the solar budget from last year. What kind of effect do you think you've had? He's really like all the rest, isn't he. Things haven't really changed."

I replied that I think I'm in a position to do something, to have some effect. "I've got to do it while I have the opportunity," I told her. But she persisted:

"Was what you did worth one month out of your life—out of our lives? Last year you said the same thing—that the three months of work on the National Energy Plan were worth it. But that was a horrible period—nights, weekends, to Washington three times a week, calls here at home. The kids didn't know if they had a father. What kind of a life is that?"

My own feeling had been that I have to do something even when I can't be sure it will change anything. I have to try, and I told her so.

"But what is it costing you?" she asked me. "Do you consider the effects on you? On us?"

I do consider what my energy work is doing to us all. I feel I can't control this work. But I realize also that I can't turn to my wife or my daughter or son and suddenly say, "Now I'm ready, I have some time now. Are you ready?"

My wife reminds me that I told her some years ago about another associate who had justified work for a terrible client by saying he could be a good influence on that client. My cause may be different, she says, but my words of justification sound much the same to her. What, she asks—and I ask myself, do my speeches, papers, reports mean to others or most important of all, to me, if they are not consistent with my life?

We understand these feelings in each other, for we have learned to talk about these doubts and concerns as well as about our love. But, I tell my wife, there must be some way I can balance my work and our family life. And she tells me of the contradictions in myself. In one week recently I flew to the West Coast to give a speech on energy conservation, the next week I was in Europe to speak on solar energy. We both know that not only have I been away from home all that time, but what kind of statement is it for energy conservation to spend all those hours of flying on jet planes in order to speak of these things for two hours? I did it. But more and more we wonder at such contradictions as they affect our family, my work, and our country.

Questions

—To what extent are we held in a high-energy way of life by our own and others habits and expectations? What else holds us in these ways of doing things?

—What is the role in energy discussions of "technological optimism" and "technological pessimism"? Which are you?

—Using this vignette as an example, would you judge that high-energy consumption has brought happiness to the American family? To the American male? Are patterns of over-work and neglect of family frequently associated with high-energy consumption work patterns?

Women in an Energy-Intensive Life
by Elizabeth Dodson Gray

"Energy has removed the drudgery from women's chores and brought the 'good life' to American women." The man was talking on and on like an ad for electricity. The enumeration of what constituted "High Quality of Life" included such phrases as Healthy Biosphere (air, water, food, etc.); Growth of Persons; Useful Work; Personal Relationships, Sense of Identity; and Power To Decide About One's Own Life.

Suddenly I was angered. It was complacently assumed that that American women had been brought to the Promised Land by providing us with an abundance of gadgets. "Listen," I burst out, "if anyone should have all those good things, I should. 'But I don't. I'm affluent, I'm white, I'm well-educated, I'm forty-eight, I'm married and with two children." Then I started to tick off the items on his list.

"Take your clean air and clean water and healthy food. I don't have those!" The air was so bad one day last summer at our place at the ocean in Connecticut that my teen-age children got chest pains after just horsing around in the water and breathing that air. I can buy a summer place at the ocean—but I can't "buy" the fresh air to go with it!

Or water—my water is full of chlorine put in to counteract what else is in it. Or my food—in processing, more than five thousand chemicals have been added, not to mention the herbicides and pesticides on what I quaintly used to consider were "fresh" fruits and vegetables. There are lead and mercury in my seafood, hormones added to my beef and chicken, and probably now even PCBs in my own breastmilk! As Betty Furness said, "I am supposed to be the beneficiary of this food system, but I am beginning to feel like the victim!"

That's exactly it, because the worst part of my situation is that I am helpless to do anything about my air or water or food. I do not have this power over my own life. That power to pollute my air, my water, my food, is held by faceless men in glass-enclosed high-rise office buildings—who decide on the basis of what is "efficient" and "profitable" for them, not on the basis of what is necessarily healthy for me.

Oh, there have been blessings. I do not fear freezing in winter as our ancestors used to. I do not spend all day washing clothes, or hand-beating rugs. And I eat oranges and lettuce brought in by truck and jet plane. But it is important for you to see that the same high-energy lifestyle which has taken from me the drudgery in those daily chores and allowed me to eat oranges, has locked me into other traps fully as disastrous for my personal growth or sense of identity.

The same fossil fuel that made cars, commuting and suburban living possible has locked me, the American suburban woman, in that isolated suburban home and nuclear family where I am expected to buy for it, clean it, maintain it, and rear children in it. It is convenient for the economic system that I do this—"convenient social virtue" is what J.K. Galbraith called it—because that same high-energy culture that exists and makes money on the voluminous production of things (cars, widgets, what-have-you) must have me, the American woman, to buy and maintain those things.

So if the American man is locked in producing by our high-energy culture, I am equally locked into consuming, urged ever onward by the sweet goad of advertising to feel that yesterday's luxuries are today's necessities. I am told by this consumer culture that my identity is defined by what I have—house, car, stylish clothes, the latest appliances. I am supposed to feel good about myself if I have lots of expensive things.

I live in suburbs made possible by fossil-fueled cars. "Marvelous sense of freedom, power and mobility" the car ads say. But do they know the way I must dribble away my days ferrying supplies from stores and cleaners to suburban home, or carrying children in endless trips to orthodontist, Scouts, piano lessons, and so on? That is not freedom. As for my identity, it is that of an unpaid chauffeur or taxi service.

The truth is that over the past 100 years the increasingly energy-intensive economic system in the United States has removed all of the production functions from the home and family. It has created a "real world" of paid work and economic and social power for men, and has left the woman to tend an increasingly empty shell known as "the home"—a rest-and-renewal "pit stop" for men who have their real being in the real world. I am to remove the dirt, diaper and nurture the young who are not yet ready to fly out to the real world, and remain sexually attractive while I use my "spare time" in socially useful volunteer activities.

I look at countless friends who have lived as I have described and now in their forties find themselves divorced by those "real world" husbands in mid-life crisis who now prefer younger "chicks." My friends who had defined their total identity as somebody's wife and mother, suddenly now are being discarded into aloneness and minimal finances, told to "find themselves" as they are thrust out into a "real working world" which labels what they have been doing for twenty years as nothing anyone wants you to list on a job application. That does not constitute "quality of life"!

"How can you blame energy for the rising divorce rate?" you may ask. Fossil fuels have been the concentrated energy sources which have powered the whole growth system. It is the accelerating pace of economic achievement in this economic growth system which drives the lives of the men in it faster and further away emotionally from that backwash of the home. Then you understand that energy has been basic to our creating these two sharply defined worlds of work and home. Marriage, like the ancient Colossus of Rhodes, has tried to keep one foot in each world. With each passing year that has become more impossible to do.

No, we must face up to the reality that our high-energy systems have brought to the American scene not only bright and shiny "things" like new cars, sleek office buildings, and affluent suburbs, but it has also brought us the seamy realities of abandoned inner cities and abandoned marriages.

Sometimes I think of St. Paul's phrase about struggling not with people but with "principalities and powers" when I look at those sleek office buildings filled with faceless men dressed in three-piece suits making decisions about my life without consulting me. I can't get at them; they are not elected by me or by anyone. Yet when they decide to have less mass transit and to finance more automobiles and highways, they make a decision for my lungs which for me is an irreversible decision. If they choose to finance nuclear power instead of the development of solar and wind energy resources, my children and my children's children will live in a world with the radioactive wastes. Yet nobody asked me—the one who bore those children and and gave twenty years of my life to rearing those children. Nobody asked me! The power to decide resides in those buildings, and I am supposed dutifully to feed my children into the maw of whatever system "they" create. And not cause trouble. As a woman, that is supposed to be my role in life.

Now just at this moment when I realize how oppressed and impotent I am before these men, these wielders of power over my life, I suddenly remember that I am also one who wields power over other people's lives. This is amazing, contradictory, and yet true.

Because I am white, I participate in the vast and subtle network of institutionalized racism in our culture. Because I am affluent, I live near the top of a pyramid of economic power in which the top 20% of the U. S. population owns 80% of everything which can be owned privately. Because I am an American, I belong to the 6% of the world's population who each year consume 30% of the energy and resources the world produces—and we produce the same share of the waste and pollution. As I walk, through the days of my normal American life-style I am using up resources which are the heritage of the people of the entire earth, but which I can get because I can "afford" to buy them.

So those faceless "principalities and powers" are not somewhere else as if apart from me. As Pogo says, "We have met the enemy, and he is us!" Our energy-intensive lives are all very interconnected and I play my own part in that faceless power which oppresses others. Dr. Seuss in one of his children's books writes about Yertle the Turtle. Yertle is the top turtle. He is held up into place by countless other turtles standing on the backs of still other turtles. I am one of those turtles. I am holding up those towers of power in my culture. I am holding them up by my decisions as a consumer and by my silent consent to the status-quo systems of institutionalized racism, sexism and American affluence in a world of poverty. I am the one who is still addressed by the prophet Amos' words of "Woe to those who lie upon beds of ivory, and stretch themselves upon their couches, and eat lambs from the flock, and calves from the midst of the stall" (Amos 6:4).

When I recognize that my energy affluence is purchased with the resources of other people, and with the destruction of strip-mined lands and miners' black-lung disease, and with radioactive nuclear wastes which must be isolated from all life for 24,500 years—recognizing all that is a start. Perhaps then I will begin to associate *repentance* in our day with lowering my taste for affluence. Perhaps then I will see Quality of Life as meaning for me "simpler living." Perhaps then I will hear anew Jesus' words from the Sermon on the Mount about not laying up for yourselves treasures on earth and about the impossibility of serving both God and mammon (Matt. 6:19, 24). With each consumer decision I will realize I give power to—or take power from—those faceless men, those principalities and powers, at the top of our economic pyramids. I have the power to buy, to forego, or to substitute for their goods—to choose simpler alternatives. I can choose my smaller car, my better-insulated home, and my well-pruned and less-burdened-with-things life. I can even join others in consumer-action and environmental action groups to demand that all who are affected by energy decisions be consulted about them.

Do you recall how the story of *Yertle the Turtle* ends? Mack, the plain little turtle at the bottom, "He burped! And his burp shook the throne of the king!" I am beginning to feel that economists have been telling us a great truth—that the economy is guided by the multitude of small consumer decisions that are like plain little Mack; they are holding the entire system up. Perhaps it is time for consumers to burp.

Questions

—Would you judge that, when all things have been considered, a high-energy lifestyle has brought greater costs or greater benefits to the American woman?

—Do you believe we would have made our decisions somewhat differently if the costs and benefits of high energy lifestyles were known more clearly in advance?

—What difference does it make to you to recognize that you are oppressed by aspects of a high-energy lifestyle and that you are also an oppressor of others by living that lifestyle? Did you realize that the same person could simultaneously be both oppressor and oppressed? Why is this so?

"DO YOU REALLY THINK WHAT WE DO CAN SAVE THE WORLD?" / *by Scott Paradise*

"Either you had better change the way you talk—or we had better change the way you live," my spouse suggested after she had heard me speak to a church group about the impact of the American life style on the world food and energy crisis. I had concluded with the suggestion that the American people ought to adopt a lower-energy life style as an ingredient toward the resolution of the crises.

Not long thereafter, one of my teen-age sons asked at supper, "Do you really think what we do can save the world?" "I think it's better to try to be part of the solution than to be part of the problem," I responded evasively.

I knew he was right. Obviously no single family can even slightly relieve the world's crisis by living more simply. Even the most Herculean self-denial would disappear in the immensity of the problem. Even more than that, if several million Americans began to live more simply, what they did *not* consume might be quickly gobbled up by those who preferred to live otherwise. Even if *all* Americans decided to live more simply, this act—without appropriate political and economic policy changes—might merely bring on a world economic depression.

But as long as the American people generally continue to seek to increase still more their energy consumption, we continue to offer the whole world a pattern of the good life for the rest to emulate—and the energy crisis becomes increasingly difficult to resolve.

As a practical question, I asked myself how, if *I* felt that happiness required a high-energy life style and indulged in it myself, how could I expect anyone else voluntarily to reduce their own energy consumption? On second thought, however, I felt it might prove possible to reduce our energy use without reducing our quality of life. And so we have done some things and have proven, to a degree perhaps, this point.

We graduated from a two-car to a one-car family. Twenty years ago we had a full-sized Ford. Now we drive a Honda Civic. When we moved to Boston we bought a large Victorian house. After ten years of rattling noisily and expensively around in it, we remodeled it, moved to the third floor, and now rent the house below with its ten spacious rooms and cavernous halls. We spent some weekends putting inches of fiberglass insulation under the roof and eaves. We spent plenty paying for storm windows and doors. Our dishwasher is now long-gone and we take turns washing dishes by hand.

These changes have sometimes caused inconveniences, but we are anything but martyrs. In fact, every change has brought with it compensations.

Getting rid of the second car, I admit, was not easy. In earlier years before we bought our second car, I as the prime breadwinner had had first claim on the station wagon, and my spouse as the prime breadbuyer had to take it when she could get it. The second car gave her a sense of freedom and independence that she did not want to lose. As "a car in being" sitting in the driveway, it meant to her she could, if she wanted, take off. My suggestion that we do without it led to a strained discussion and only won acceptance with my assurance that we were not returning her to her previous condition of servitude.

Now she has a job that she cannot easily reach by public transportation. So except when I have special needs, she lays claim to the car for her commute. In winter I brave the elements, wait for the bus, trudge around the city to my appointments. But then, I don't sit in traffic jams, circle endlessly searching for parking, or wait to be skinned at the car repair shop. Most days during eight months of the year I bicycle the eight miles to my office, breath fresh air, keep in shape, and come home sweaty, tired, relaxed, and happy.

Four of us living in our third-floor loft apartment means less privacy, more irritation over loud rock on the stereo, and frequent distraction over the endless phone calls for other people. It means less space for dinner parties, and no space for any but the most familiar overnight guests. But it also means less housework. It makes unavoidable relationship of a sort, or at least bumping into each other. It has also proven to be fun.

And once installed or taken for granted, storm windows, insulation, lowered thermostats, and the dishpan become all but invisible.

The pain comes from realizing what we haven't done. The boys, of course, take the brunt of the blame. It is they who take fifteen-minute showers. It is they who forget and leave the oven on all night. Meanwhile, after railing against such energy extravagance, my spouse and I think nothing of flying all over the country to consult, or to lecture about energy conservation. And between us, we put 16,000 miles each year on our Honda Civic.

Of course it is not clear how far we could—or should—go toward a lower energy-use life style. Everything done suggests something further which could also be done. And when these things are done, we still consume many times the amount of energy available for the average citizen in a Third World country.

Moreover, our efforts are limited to a marked degree by the system of long-term capital investments that have been made in the past which presume a high-energy life style. Land-use patterns, for instance, have developed so that life without a car would be inconceivable for most of us. Public transportation has so deteriorated that for many families two or three cars seem necessities more than luxuries. The architecture and landscaping of many houses are such that even with insulation they will be cold in winter without heavy inputs of energy for heating.

To reshape our land-use patterns, to redesign our transportation system, and to reform our architecture will take public action over a generation—and we cannot but feel we are frustrated victims of the decisions of the past.

But doing something about our personal life style has perhaps reduced our frustration. For us the experience I've described of starting down the path toward more simple living has *not* been sacrificial. Our earlier life-style decisions, like those of most Americans, were made to avoid minor daily inconveniences. But the more we have cluttered our lives with gadgets, chemicals and services, the more the inconveniences grew back again, hydra-like, in endlessly proliferating forms. The telephone, for instance, serves as a great convenience. And yet it means anyone can interrupt our dinner at their convenience, not ours. An alternative path to convenient living grows from simplifying life and learning to be careful about little things. We can develop the habit of asking ourselves, Will consuming this or acquiring that *really* improve my life experience? Trivia sloughed off may make more room for what may be more significant satisfaction and fulfillment.

If the energy crisis is not resolved, many of those now living on high-energy budgets may find they *have* to learn to live more simply. If enough of us learn the satisfaction of more-simple living soon enough, some aspects of the energy crisis may be averted. But the wisdom of finding fulfillment in living simply long predates the energy crisis. It may be something of this wisdom Jesus had when he said, "A person's life does not consist in the abundance of possessions."

Questions
—Would you find painful the sort of changes this family made?
—How do you feel about making lifestyle changes toward using less energy, even though such changes are such a tiny part of the overall energy problem?

Do you think such changes and energy savings might be worth making anyway? Why?
—How far can we go toward setting ethical standards for a whole family to live by? (How far can a person legitimately go toward coercing others in the same family to live by certain ethical standards?)

"I DON'T WANT TO BE KILLED NEXT SUMMER"
by David Dinsmore Comey

I am bothered by writing this vignette. I generally resist probes of my thinking. I once told a nosy Congressional committee, "I was born an Anglican and a Republican" in the hopes it would shut them off. But it didn't.

I contribute to the energy problem in many ways, primarily through the consumption of kerosene jet fuel. I am president of a corporation whose headquarters are in Chicago but I travel a lot to our other offices on both coasts and also to Europe and the Middle East. I sit on the board of directors of four other corporations, and also on a number of panels and commissions that advise the Congress and the President. Between the jet fuel and all the paper I cause to be generated, I am responsible for the expenditure of many millions of BTU's per year.

I use the telephone whenever I can. But people seem to think there is something magic about sitting across the table in person, and they are willing to spend enormous amounts of money and time to do things that way instead of relying on the telephone or Telex. It really is quite silly when you think about it.

I got rid of my car eight years ago and have never missed it. I try to walk to work when it is good weather: it gives me an hour to think, away from the telephone. I also like riding the bus: it gives me a chance to talk to people and find out what is on their minds. (Actually, I do have other, more scientific channels inasmuch as our company conducts interviews with over a million households a year.)

I startle a lot of people on the bus when I ask them their views, probably because of the way I dress. My children accuse me of having been born in a three-piece Brooks Brothers suit, complete with my gold watch. They don't understand why I feel uncomfortable wearing anything else, except on Sundays, when I wear a morning coat and striped pants. The dean of the cathedral doesn't like it—and then I don't care much for his new prayer book either.

It is mostly a matter of my upbringing: Social Register, lots of servants, New England prep school, Ivy League university, and all that. I am at least less formal than one of my uncles, who was a bishop so High Church that he made all the priests in his diocese wear cassocks all the time!

I don't like change. I like cathedrals: Canterbury and Salisbury best of all. I make an annual pilgrimage to them to stay in touch with my roots.

But I cannot ignore the changes going on right now, and the more drastic ones yet to come. Virtually all of them are directly connected to our energy use. Our national debt is soaring. The dollar is dropping because of oil imports. We are like spendthrifts consuming out of principal, when what we need is a prudent trust officer to keep us living within our income. The principal we must live off is the finite set of natural resources and fossil fuels that we are depleting so rapidly that we will leave nothing for our beneficiaries of the next generation.

We could live very well on our energy income—if we were willing to use the next thirty to fifty years to convert to renewable energy sources that rely on natural flows of sunlight and wind and wastes from our other natural activities. We would then be truly independent of foreign sources of fuel, because all our energy would come from decentralized sources distributed wherever the energy was needed in the country.

Such an energy policy would also go a long way toward solving the problems of the people who ride the bus with me. Those who are lucky enough to have jobs are paying an enormous percentage of their budget for fuel bills, because their buildings have not been made energy efficient and the cost of gas and electricity keeps going up.

Our local lighting company's solution is to build more nuclear plants. They already have the largest number of nuclear plants in the country, and they plan to double that in the next ten years. Their argument is that without the energy from those nuclear plants, the urban poor will be forever mired in poverty. Some of the physicists and economists on my staff have looked at these arguments put forth by the nuclear industry and found them to be without merit. I will try to confine myself to just one issue among the many they found: the issue of jobs.

Many of the people riding on the bus are looking for jobs. But they cannot find them. Here in Chicago, unemployment among black males under 35 is so bad that only one in six has a job. In Detroit it is one in seven. Every source of information coming to me is saying that if there are not riots in the summer of 1979, they will surely come in the summer of 1980. And regardless of the timing, the severity will make 1965 and 1968 seem like a kindergarten party.

In other parts of the country Hispanics and other urban groups suffer similar problems of unemployment. Yet our national energy policy and our jobs policy are not being coordinated to solve these problems simultaneously.

Building nuclear plants, oil refineries and other energy production facilities is the worst possible way to increase employment. Why? Because they require an investment of several hundred thousand dollars for each job created. When compared with the approximately $20,000 required to create most manufacturing jobs, nuclear plants eat up our precious limited supply of capital, contribute to inflation, and deprive us of productive jobs in the remainder of our industrial sector. Their construction will exacerbate our unemployment problems, not solve them.

By comparison, the construction of small, community-sized solar energy systems generates four times as many jobs as an equivalent nuclear powered system. Not only does solar energy mean far more jobs, but more importantly it means jobs which use skills that the people who are currently unemployed already have or can easily acquire. These include carpentry, sheet-metal working, glazing, plumbing, wiring, and so on. To get a job working for a solar installation company, you don't need a Ph.D. in nuclear engineering, nor do you have to spend years learning how to weld zirconium alloys.

The capital requirements are also on an entirely different scale. It now takes over two billion dollars to build a single nuclear plant. Even wealthy utilities are having major problems raising that kind of capital in the money markets. By contrast, you can start a small contracting firm to install energy-conserving insulation and solar heating systems with a loan from the Small Business Administration.

Once you begin employing people from the local community, all sorts of interesting things start to happen. Because the jobs are where people live, the traditional problem of finding transportation to get to work is solved. The jobs are also highly visible in the community, unlike jobs on an assembly line fifteen miles away. Younger children can see the work being done. It is a better role model for them than the customary pimps and pushers. The jobs are satisfying too, for the end product is not a piece of some distant project, but is something built with one's own hands that is a permanent improvement to one's own community. Hence it is not likely to be vandalized. What was once ugly and dilapidated is transformed into something livable and pleasing. And the money spent stays in the local community and increases the flow of goods and services there. The local economy stabilizes and begins to become more self-sufficient.

The technologies involved also lend themselves to community decision-making. You need not call in radiation experts in order to discuss the hazards of passive versus active solar hot water systems. And if you are dissatisifed with the service of your new district heating facility, you can go complain to the manager in person—instead of having to communicate with the omnipotent computer at the power company. You begin to feel that what you do can make a difference—instead of feeling like a cog that can be manipulated without anyone caring.

Proper energy conservation and our use of renewable energy systems can reduce our dependence upon traditional fuels to the point where people will be in control of virtually all of their own energy supply. People need not be subject to the capriciousness of computers, sheiks, and bureaucrats. Local self-reliance is one of the things that made this country great, and it ought to be encouraged once again.

Fortunately some of my friends in the investment banking business are starting to understand this. It has helped, of course, when their own studies showed that the average monthly fuel bills would by 1983 exceed the mortgage payments on most of the paper [i.e., mortgages] they held, and that they might be stuck with a lot of bad paper unless they encouraged energy conservation and renewable sources like solar energy.

But the U.S. Department of Energy is still pushing large-scale energy facilities which are centralized and capital-intensive and which will run on rapidly depleting fuels (including uranium) and require severe impacts on the environment and human health.

There are not very many opportunities that come along in which two separate major public policy problems—energy and unemployment—can be solved by a cost-effective and achievable program. I believe that use of renewable energy technologies can help solve our critical urban social problems. I could buttress my case by further recitations of technical advantages, moral and ethical considerations, arguments about stewardship and responsibilities to my neighbors. But my real reason is one of pure self-interest: I do not want to be killed during the summer of 1979 or 1980, and I do not want other people to die either.

Questions

—Do you find the strategy for dealing with the urban unemployment and energy needs convincing? Why? Or why not?

—Do you find that strategy appealing? Why? Or why not?

—Do you see any support for this sort of strategy in our religious tradition?

A COMMUTE TO THE AIRPORT
by Bolton Williams

It was a cold bleak day but my station wagon was warm and comfortable. The rush-hour traffic was just beginning to build on the expressway and was flowing smoothly between 65 and 70 miles per hour. No one was worrying about the 55 mph speed limit and the flow of traffic was not yet heavy enough to force anyone to slow down appreciably.

I took the exit which would place me on a back road to the airport. I was quickly out of the city and crossing a five or six mile stretch of undeveloped country. I was looking forward to the openness and the trees. It had been a good while since I had passed this way, and I was surprised to see that in the interim there had been considerable construction activity. Midway through the stretch a right-of-way had been cut for another branch of the expressway system, and it was obvious that the surrounding areas had been surveyed and the inevitable development process had already begun. I had a sinking feeling as I contemplated the loss of this pleasant interlude in my future trips to the airport.

As I neared the vast airport complex, activity increased dramatically. The six lanes of highway into the airport was congested with traffic and the air was filled with arriving and departing flights. Motels and restaurants, freight-handling facilities and rental-vehicle agencies filled an area which ten years before had been pasture-land for cattle. I watched the outgoing flights as they soared from the runways and rapidly disappeared into the low overcast, each of them leaving behind a pall of exhaust smoke to record their passage. I parked the station wagon and went inside to await my own "silver bird" and its soaring flight into the brilliant sunshine above the overcast.

Our modern society has provided us with a mobility undreamed of fifty years ago. We control our environment with air conditioners, heating systems, humidifiers, de-humidifiers, dust filters. We are informed of happenings throughout the world almost as they happen. We have microwave ovens to warm our food fast and crock pots to cook it slow. We have machines which wash and dry our clothing, wash and dry our dishes, grind and compress our garbage, crack our ice cubes, mix and blend our food, clean and scrub and brush and wax our floors, answer our telephones in our absence, and we even have electric carving knives for the Thanksgiving turkey. It is truly a great age in which to live. What difference does it make that gasoline to run the station wagon has doubled in price in the past five years and that our gas and electric bills increase monthly. Look what we have!

But what do we have—in addition to our great mobility and convenience and communications?

We have air polluted by the fumes of our vehicles and by the factories which produce them and by refineries which fuel them. Our waterways are fouled by chemical and thermal and sewage effluents. The lands beyond our cities are scarred with landfills to dispose of our offal and with mines to provide our raw energy sources. Even the oceans show the blight of our extravagant society, as oil spills and sewage-sludge dumps foul the waters. In ever-increasing numbers we scurry from our cities into the open spaces, seeking nature and solitude, a release from the pressures of crowding, of time schedules, and "progress." We seek recreation.

Should we not be able to find some significance in this? I am familiar as a biologist with the ecological concept of "carrying capacity." Quite simply, this involves the premise that a given area can support a certain population-size. This population must contain organisms which produce food, generally plants which convert carbon dioxide and water into carbohydrates. This same population must also include consumers and decomposers which utilize the carbohydrates and return them to a form reusable by the plants. A balance is developed in nature between these components, and populations are maintained (and limited) accordingly. Should any of these components grossly exceed these population limits however, the whole system breaks down and producer, consumer and composer components do not survive.

Our technology has permitted us to circumvent this carrying-capacity principle in many ways. One American farmer can produce enough food for fifty people from a plot of land which at the turn of the century would barely support his family. But he does this only through his use of additional energy.

In many areas of our land we exceed the carrying capacity by exploitation of fossil-fuel energy sources and also by importing resources from other areas which are thus maintained at lower populations. With this fudge-factor on the carrying capacity of our local and even national environments, we have succeeded in developing urban populations far beyond the natural carrying capacity of those areas.

We have also become accustomed to our machines and to all our other devices which also require energy. And we have ignored or overlooked the accumulating strain all these place upon the ecosystems around us and upon the global biosphere itself. We do not want to have to give up these aids and conveniences. What difference does it make to a suburban homeowner or an urban apartment dweller in Los Angeles (or in Chicago or Houston or Atlanta or Philadelphia) if a Montana ranch or an Appalachian valley must be sacrificed to provide energy to power their city and if a great gaping hole must be torn in the earth to get at the fossil fuel which will fire a great electric-generating facility?

Have we in our comfortable modern existence lost touch with the reverence the Indians held for the land? Can we possibly believe that man can go on forever raping the biosphere given to us by our Creator, without eventually having to pay the consequences?

Like everyone else in our society, I too have become fat and lazy and dependent upon our technology. I see fields of concrete which once there were grasslands producing oxygen and food. I see polluted streams and rivers which once teemed with fish. I see a gray-brown haze over our cities rather than clean air and azure-blue skies. And yet I must struggle with the thought of what I—and we—must give up if these things are to change.

Questions

—How do you feel about the environmental description in this vignette?

—Are you prepared to change in order to reduce our adverse impacts upon the human environment? How?

—Do you think that we should take seriously the warning implicit in the biologist's discussion of "carrying capacity"?

Appendix B

Energy in the News

WHY WE MUST ACT NOW
Reprinted from Newsweek, July 16, 1979.

The problem is as grave as any America has faced since World War II—a national-security issue of such scope and complexity that it is transforming the economy, threatening to destroy political consensus and undermining traditional life-styles. Jeremiahs speak darkly of an inflation-plagued, no-growth economy for a generation to come or, alternately, of the necessity for grim personal sacrifice by American consumers. Fuel shortages pit East against West, city against suburb, farmers against truckers in a regional divisiveness not seen since the issue of race split the nation North and South. Yet, for all the ferment and warnings of the past half-dozen years, America has done little to solve an energy crisis that threatens to engulf it.

The propensity of the nation to search for villains—Middle Eastern sheiks, oil-company executives, environmentalists and government bureaucrats—only serves to paralyze it and prevent proper action. America runs on oil, and its problem is largely its dependence on an uncertain source: what President Carter has correctly called "a thin line of oil tankers" stretching from New York and Houston to the Persian Gulf, one of the world's most politically unstable regions. It is also a crisis of price: a nation accustomed to cheap as well as abundant oil has yet to adjust to the new era. Economist Thomas C. Schelling, director of Harvard University's graduate school of public policy, accurately describes the new American challenge. "The problem now," Schelling says, "is to meet the rising economic cost of fuel with policies that minimize the burdens, allocate them equitably, avoid disruptions in the economy and keep the cost from rising more than necessary."

Massive Tax
In its failure to deal with the problem, America has already paid a terrible price. In less than a decade, the nation's bill for imported oil has increased from $1.3 billion to $60 billion—a massive annual tax on purchasing power

Copyright 1979 by Newsweek, Inc. All rights reserved.

that has reduced economic growth and spurred inflation, sent the balance-of-payments deficit into orbit and subjected the U.S. dollar to regular and humiliating depreciation. If nothing is done, America will import 12 million barrels a day by 1985—50 per cent above the current level—and its import bill will amount to $10 million *an hour* at existing prices.

The U.S. hunger for imported oil has also levied an indirect toll on other countries with little or no fossil-fuel reserves of their own. Developing countries, in particular, have suffered even more than the West every time the OPEC cartel has raised the world price—and America's unsatisfied appetite for foreign crude has enabled the higher price to stick. As a result, growth rates in most Third World counties have been depressed, and even worse, the buildup of debt in developing countries to pay for their oil imports has soared to dangerously high levels that have sent tremors through the international banking system.

America's growing dependence on imported oil leaves it at the mercy of internal events half a world away. Iran's winter of discontent led to the American summer of scarcity, even though the shortfall wasn't very great—at most, perhaps 5 per cent of total U.S. oil consumption. The Iranian experience has also been a warning to other OPEC nations that rapid development based on oil revenues can lead to social instability. The result may be that Saudi Arabia and other less-populated oil producers will limit their oil output. In a world of scarcity, the realization that oil left in the ground can only appreciate in value means that the OPEC countries will never produce as much as the U.S. wants to consume.

Oil Linkage

The dependence on foreign oil also leaves the U.S. vulnerable to political blackmail. Nigeria says that it might cut off the 1 million barrels of oil a day it sends to America if the U.S. removes its economic sanctions against Rhodesia. Libya's quixotic leader, Muammar Kaddafi, hints that he, too, might cut sales because of U.S. policy toward Israel. Even "moderate" Saudi Arabian oil minister Ahmad Zaki Yamani says that his country might link oil sales to American pressure on Israel to negotiate with the Palestinians.

At home, the scramble for fuel supplies has nurtured a growing divisiveness. The interests of the oil-and-gas-producing states of the Sunbelt are at odds with the needs of the big consuming states of the Snowbelt—giving rise to bumper stickers that proclaim "Let the Bastards Freeze." The deep coal mines of the East compete with the strip mines of the West. Refining more heating oil for the chilly Northeast means less gasoline for the auto-dependent West. More diesel fuel for farmers means less for truckers. And consumer groups on both coasts besiege Washington with their complaints and charges of inequity. "There is a certain energy McCarthyism at work now," says MIT economist M.A. Adelman. The result: political consensus on how to deal with the energy crunch has been impossible to achieve.

Nickel Barrels

That Americans have found the new reality of expensive, scarce fuel difficult to accept is hardly surprising,. U.S. postwar prosperity was built on cheap, plentiful energy—so cheap, in fact, that domestic oil producers persuaded Congress to impose import quotas in the 1950s to keep out Middle Eastern crude that could be produced for literally pennies a barrel. The result has been a mobile society with two cars in every suburban garage and electricity rates so negligible that the real price adjusted for inflation actually fell during the 1950s and 1960s.

OPEC ended the era in 1973–1974 when it quadrupled prices and, for the first time, restrained supply. The move came at a critical moment: America's own oil fields were showing signs of wear, with both reserves and actual production declining. But the U.S. compounded the problem with the wrong response: in an attempt to insulate consumers, Washington strengthened price controls on domestic crude oil. The controls have had a perverse effect. They have, in fact, subsidized greater use of oil by blunting the economic incentive to adjust consumption patterns through conservation. And controls have subsidized imports as well: with prices on U.S. oil output held below the world market, domestic producers saw little reason to invest heavily in a search for new oil. U.S. production has thus continued to slip—and OPEC imports have continued to grow.

To cope with chronic gaps between supply and demand of crude oil and its refined products, America has relied principally on a government allocation system that has only made matters worse. For months, the attempts to allocate scarcity by bureaucratic fiat have turned a relatively minor short fall into long lines at the gasoline pump and pleas for more equitable distribution of supplies.

Coal Paralysis

If the U.S. had viable alternatives to oil, the problem might gradually disappear. But natural-gas price controls, first instituted in 1954, still depress production of that clean-burning fuel—and neither coal nor nuclear power has picked up the slack. America has hundreds of years of proven coal reserves—the U.S. is "the Saudi Arabia of coal"—but the industy is plagued with problems that block expansion, from legitimate safety and environmental concerns to a history of labor-management disputes. Nuclear power, which once held the promise of electric power so cheap it wouldn't pay to meter it, has suffered setback after setback, culminating earlier this year in the accident at Three Mile Island in Pennsylvania and a de facto moratorium on new plant construction.

Thus, the accumulating impediments to coherent action have been building—and so have the prospective penalities of further delay. If the U.S. does not begin now to develop a rational energy policy, its dependence on OPEC oil will grow and its capacity to satisfy its own need will decline. Its economy will be plagued by chronic inflationary recessions as the Mideast sheikdoms exact their toll on the world—and sorely needed capital investment seems certain to fall short. Perhaps most important, America's political will and spirit will erode, pitting citizen against citizen in an age of energy scarcity.

But America's energy future need not be a prisoner of its energy past. The nation has the capacity to mobilize its resources to ease the energy crisis.

CAPSULE UPDATES ON STATE OF THE ART . . .
Reprinted from *The Christian Science Monitor,* December 17, 1979.

Oil and Gas

Oil and gas, which accounted for 74 percent of United States energy use in 1978, represent only 1 percent of national energy resources. Even if there were no energy supply shortage brought on by overdependence on foreign oil at this time, such an imbalanced use of energy sources would be questionable.

Not only are domestic reservoirs of gas and oil finite, but the bulk of these resources appear to have been discovered already. Discovery rates of new gas and oil fields have fallen drastically in recent decades. Noting this in a recent analysis, David H. Root and Lawrence J. Drew of the US. Geological Survey (USGS) showed that this is due to basic geological reasons, not to inefficient exploration or lack of "financial incentives."

Because of the nature of the oil- and gas-containing formations, they believe that the big fields hold most of the resources and, in the United States, these have already been discovered. New exploration can only turn up relatively small (albeit significant) supplementary sources. This assessment is clearer for oil than for gas because of greater uncertainties about gas reserves and recoverability. Nevertheless, gas and oil are lumped together in the analysis and reported in terms of equivalent barrels of oil.

They explain: "Of the more than 20,000 oil fields that have been discovered in the US, only 275 are expected to produce more than 100 million barrels of crude oil each. . . . At the end of 1975, these largest 275 fields had produced 66.5 billion barrels of crude oil — 59 percent of all crude oil produced in the US to that date — and contained 90 percent of all the proved reserves in the US. Two hundred and twenty-three of these fields (81 percent) were discovered before 1950, which is approximately the time when the high rate of crude oil discovery ended. . . ."

What's true of the US is true of the world. Half of all the recoverable crude oil discovered in the world is contained in 33 large fields. As a result the world oil-discovery rate may drop dramatically as the large, easy-to-find fields are located in the near future.

Thus the US could not rely for long on imported oil (or liquid natural gas) to meet a large share of its energy needs, even if this were considered a wise policy. In 1979 the US was using oil at an average rate of 18.4 million barrels

© 1979 The Christian Science Publishing Society. All rights reserved.

a day (mbd) of which 8.2 mbd were imported. Natural-gas usage averaged 10.4 mbd, in terms of the equivalent amount of oil, of which only 0.6 mbd were imported. There is no way that this level of usage can be sustained, although some substantial use will continue into the next century. Even imports could not be counted on to make up shortfalls in domestic production.

Thus the bottom line on the gas and oil balance sheet is a stark warning to continue exploration by all means, but to work aggressively to shift the bulk of energy supplies to a balanced mix of America's alternative sources, including heavy emphasis on conservation.

'Synfuels'

One of the most straightforward ways to quench America's thirst for fuel is to produce it synthetically from domestic energy resources. The United States has 400 to 600 billion barrels of extractable shale oil, a liquid similar to light crude oil. Domestic reserves of coal are much larger — the energy equivalent of 14 trillion barrels of oil — and there are a number of industrial processes for converting it into gaseous or liquid fuels.

"Shale oil requires little technology development, but environmental problems continue to loom large. Coal liquefaction is expensive and likely to remain so," summarized the most recent Ford Foundation study, "Energy: The Next Twenty Years."

Current cost estimates for shale oil are running at $25 to $35 per barrel, with those for coal liquids $10 to $15 higher. There is a consensus that pipeline-quality gas from coal will be cost-competitive before coal liquids. Indeed, the Federal Energy Regulatory Commission recently OK'd plans for the first commercial-size "syngas" plant, proposed by American Natural Resources. The plant, as originally planned, would be built in North Dakota, cost about $1.5 billion, and produce 125,000 cubic feet of gas per day.

The continuing crisis in Iran has created intensive congressional interest in synthetic fuels. The result of this is passsage of a $20 billion energy-security reserve to stimulate synfuel development.

The estimated cost of shale oil seems substantially less exorbitant with petroleum price increases in the last 6 months. Still, the companies involved in oil-shale development feel that loan gurantees are essential to secure investor backing for the operations, which entail $1 billion each.

The richest shale is concentrated in southeast Colorado, in the Piceance Basin, an arid plateau carved by steep-sided canyons. A 50,000-barrel-a-day plant would consume 5 to 10 million gallons of water and create 60,000 tons of spent fuel daily. According to Colorado officials there is adequate water for 10 such plants. After that, water conflicts are likely. This area is thinly populated, so building up roads, water and sewer systems, housing, schools, and other services for the fivefold increase in population anticipated would be a costly and time-consuming operation.

Coal, in contrast, is spread widely. But development of coal-based fuels competes with mining for direct burning. The commercially tested Lurgi process will work only on Western coal. It requires large amounts of water, which represents a potential problem in the semiarid areas where most Western coal is found. More-advanced methods, including converting coal into gas underground, might reduce this problem.

President Carter has set a target of 2.5 million-barrel-per-day synthetic-fuels industry by 1990. But, because of the various problems involved, many energy experts feel this is unrealistic. The Committee on Nuclear and Alternative Energy Sources of the National Academy of Sciences, for instance, estimates that, even with a national commitment, a 1.2 million-barrel-a-day capacity is all that can be expected. This could be supplemented with an annual production of 1.6 trillion cubic feet of pipeline gas per year, however.

Coal

Coal is America's most abundant fossil fuel. With known reserves that represent the energy equivalent of 13.7 trillion barrels of oil, "coal can be America's most productive domestic energy source well into the 21st century," a congressional Office of Technology Assessment report stated.

Since the Arab oil embargo, the federal government has been trying both to pull and to push industry and utilities into using more coal — so far with limited success. Since 1973, the annual growth rate for coal use has been only 1.2 percent, about half that of the previous decade.

The coal industry is caught in a complex web of problems. Burned directly, coal is a dirty fuel. Air-quality regulations have restricted its use. It is only recently that uncertainty over strip-mining rules were largely resolved by formation of the federal Office of Surface Mining. Conversion to coal by users is expensive and requires a large amount of space and extensive rail connections. The railroads are using coal to bootstrap themselves back to financial viability. Because coal haulage rates have more than doubled in the last two years, it is now cheaper for many users near the coast to import coal from Australia than to buy it from Wyoming. Historically, the coal industry has been beset with labor problems, and years of decline have weakened its management capabilities.

The failure of coal consumption to increase as had been hoped turned 1979 into one of the worst years ever for the industry. Analysts estimate there is an excess production capacity of 100 million tons a year. An unprecedented number of small operations are folding as spot market prices for steam and metallurgical coal have fallen. Coal energy is selling for the equivalent of $4 per barrel of crude oil.

Despite these problems, energy experts continue to predict dramatic increases in coal consumption by the year 2000. Projections start at about 1.5 billion tons per year, compared with 655 million tons burned in 1975, and range up to as much as 3.4 billion tons.

Such increases would be due primarily to utility use. Use of coal in industry faces severe problems. Although there are certain regions where more people are turning to coal for home heating, the general feeling is that only via district heating will coal play a significant role in the residential sector.

One way to reduce shipping costs of coal is slurry pipelines. There are eight proposals to build pipelines of this sort, which would carry finely dissolved coal suspended in water. Although generally considered more efficient than rail cars, slurry pipelines are being fought by the railroads and face local opposition in such arid Western states as Wyoming because of the large amounts of water they require.

Meanwhile, progress is being made in more efficient and smaller-scale coal power plants. An example of this is a 5-ton-per-hour plant at Georgetown University. This suspends burning coal on a "bed" of high-pressure air, resulting in more complete combustion. The economics of this system is still uncertain, but researchers hope this approach will make it possible to heat and cool large buildings at reasonable cost.

Nuclear

Although the nuclear industry is embattled, it still is a major energy supplier in the United States. Seventy-two atomic power plants supply 3.6 percent of the total energy demand, which was 276 billion kilowatt-hours in 1978.

In addition, 91 reactors are in various stages of construction scattered throughout the nation. When, and if, they are completed, they will add another 100,100 megawatts of electrical capacity — more than twice that of currently operating units. If all this electric capacity were fueled by oil, it would add 6 million barrels to the 8.2 million barrels a day the US now imports.

Still, the future of nuclear power remains in grave doubt. Its major problems are safety of reactor operations, waste disposal, and uranium supplies.

Widespread public concern about the safety of reactors was created last April by the mishap at the Three Mile Island plant. Although no one was injured, the apparent disorganization of the industry and the Nuclear Regulatory Commission (NRC) shook public confidence.

Meanwhile, the question of what to do with radioactive waste remains a vexing problem, one that is more political than technical. Hardly any one wants a radioactive dumping ground in his backyard.

Finally, there may not even be enough uranium in the US to power the reactors now under construction. The current generation of reactors uses uranium inefficiently. This is because only 0.7 percent of naturally occurring uranium actually "burns" in these reactors. A way to extract substantially more energy from this source is to transmute a large percentage of the nonburning uranium into plutonium in "breeder" reactors. This has the effect of increasing the recoverable energy from the equivalent of 120 billion barrels of oil to more than 14 trillion barrels.

However, the US breeder effort has lagged behind that of France, Britain, and the USSR. Concern over large amounts of plutonium — particularly the fact that it can be fabricated into a crude atomic bomb with relative ease — has caused the federal government to put development of the breeder reactor into limbo.

Today's nuclear reactors are not the only types possible. Fusion reactors work on a totally different principle. Their fuel is deuterium, the most abundant of "fossil fuels." There is enough deuterium easily extractable from water on Earth to meet all of mankind's energy needs indefinitely — if the nuclear fusion process that releases the energy can be harnessed economically and safely.

After a quarter-century of effort, experts still do not know whether this can be done. Fusion is the type of nuclear process that powers the sun and stars. It takes place at temperatures of many tens of million of degrees. The two leading approaches to controlling fusion in the laboratory either use laser beams to compress fuel pellets to high densities and pressures or confine hot deuterium gas with magnetic fields.

Either, or both, of these techniques is expected to demonstrate fusion on a laboratory scale within the next five years. Substantial engineering efforts must follow, making fusion an energy source for the next century.

. . . IN EACH FACET OF THE ENERGY SPECTRUM
Reprinted from *The Christian Science Monitor,* December 17, 1979.

Conservation
America's leading energy option is conservation. It could also be the cheapest measure.

This is foreshadowed by typical "payback times" of five years for home insulation, savings of more than $1 for every gallon of gasoline cut from a trip's fuel budget, and investment returns such as the $700,000 a year an American Can Company unit now saves, thanks to a $73,000 conservation program that cut its energy use by 55 percent.

It's not so much a matter of doing without as it is of using energy more efficiently. In the report "Energy Future," prepared by the Harvard Business School Energy Project, Prof. Daniel Yergin estimates that the United States could cut its energy diet by 30 to 40 percent and still maintain its living standards. This is an optimistic, but not impossible, assessment of what could be done, given public awareness of the possible savings and the national will to achieve them.

It is a formidable challenge. It means massive investment in new equipment and buildings and changing many traditional outlooks.

Consider the automobile. New cars have kept up with fuel-economy standards, which now require an average of 20 miles per (US) gallon (m.p.g.). But, as Resources for the Future noted in its study "Energy in America's Future," the average for the US auto fleet as a whole has barely begun to move up from about 14 m.p.g. It takes time to replace the existing fleet, especially when people are hanging on to their cars longer and have shown a tendency to buy vans and light trucks, which have relatively poor fuel economy. Given this fact and the difficulties automakers see in getting higher fuel economy, the

© 1979 The Christian Science Publishing Society. All rights reserved.

goal of a fleet average of 27.5 m.p.g. by 1985 seems elusive. Yet, in the long run, most experts expect to see substantial savings in gasoline consumption— for example, a drop of 29 to 48 percent in automotive energy use by the year 2000 compared with 1976, according to the Resources for the Future study.

Comparable savings are possible in buildings, manufacturing, and other areas. For example, the study foresees a drop in the growth rate of energy use by housing (now about 10 percent of national energy use), because of trends toward better insulation and more efficient furnaces already under way. Aggressive conservation measures could cut back that energy use even more.

In short, the message of the experts is that conservation can bring major energy savings, but little will happen without much more effort than is being made today. The greatest obstacle is the lack of a coherent and coordinated national policy that champions conservation as a high national priority.

"Sometimes one concludes that the real challenge of energy conservation is not to do it, but rather to believe that it can be done," Professor Yergin says.

Renewable Fuels

Biomass is a term recently coined to cover the multitude of plant and animal materials that can be used as renewable energy sources. It includes everything from wood to municipal waste, from sewage to sorghum.

Currently, biomass — primarily wood — provides Americans with more than half as much energy as that produced by the nation's nuclear reactors. Projections of its potential for the year 2000 range from 2 million to 14 million barrels per day oil equivalent.

Despite its diversity, biomass has some common characteristics as an energy source. Its energy content is lower than fossil fuels. It contains a large amount of water. It is widely dispersed. On the positive side, it is low in sulfur and, in general, produces less noxious air pollutants when burned.

These materials can be burned direct for heat or in a boiler to produce electricity. They can be converted into a gas or a liquid fuel.

Whatever the form, the current wisdom is that biomass power is more expensive than that from fossil fuels. Nevertheless, utilization is growing rapidly. Use of wood, for instance, has been increasing by 15 percent a year since 1973. In fact, demand for firewood is so high in many areas that there are not enough foresters to supply it, according to the American Forestry Institute.

Development of a mobile unit that turns trees into wood chips on the spot and a patented process for compressing chips into pellets with a heat value equal to that of coal are reducing the cost of wood heat substantially in some areas. Also, the Department of Energy is funding a 1,000 acre "energy plantation" in South Carolina which will grow trees like agricultural crops in hopes it will reduce the cost further.

Gasohol is another form of biomass utilization which is on the upswing. One part grain alcohol is mixed with nine parts unleaded gasoline. Hard-core interest is centered in the Corn Belt of the Midwest but is spreading rapidly. This year the Bureau of Alcohol, Tobacco, and Firearms expects over 5,000 applications for small distilleries, most on farms.

These distilleries convert only the sugars in grain into alcohol. But a recent advance at Purdue University increases the possibility that agricultural and forestry wastes can be converted into fuels. A pretreatment separates the cellulose from the woody substance, lignin, in plant matter. This makes it possible to convert the cellulose into alcohol.

Even further in the future, Nobel Laureate Melvin Calvin is experimenting with desert plants that produce petroleumlike substances. He optimistically predicts this plant petrol can be produced for $20 per barrel.

Meanwhile, major development efforts are focussing on ways to put the energy in municipal waste and sewage sludge to use. Because such operations can charge a fee for waste disposal the estimated price of the energy they produce is lower than other processes.

Geothermal

Geothermal energy (subterranean heat) is a major resource poised for take off.

In 1979 the US Geological Survey (USGS) estimated that, given reasonable development, it could meet something like 10 to 15 percent of US energy needs within two to three decades. The Department of Energy has a lower, but still substantial, estimate of 6 to 7 percent.

The US is already getting a token amount of electricity from geothermal power — the most notable development being that at The Geysers site near San Francisco. Here, Pacific Gas & Electric Company has roughly 700 megawatts of generating capacity. Farther south, in Imperial Valley, three 10-megawatt generating installations were nearly ready to start up, at this writing. California's geothermal energy office expects something like 200 megawatts of capacity to be on line in Imperial Valley by 1983.

That's still less than the capacity of a single average size nuclear power plant. But the ultimate potential for geothermal power theoretically is immense.

The USGS study estimates the country has something like 6,400 quads (quadrillion Btu) of it that could be put to use — an energy reserve equivalent to 1,100 billion barrels of oil.

This includes two types of geothermal resource. One is the hydrothermal convection that powers geysers. It represents 2,300 quads (equivalent to 400 billion bbl. of oil). The other type, called "geopressured," is hot water that is rich in natural gas and held under pressure in deep formations. If techniques can be developed to exploit them — which means drilling to depths of several kilometers — 400 to 4,200 quads of energy (69 billion to 724 billion bbl. of oil equivalent) may be recoverable as hot water and methane gas.

A third resource, whose potential is more speculative but which is available in all parts of the country, is hot dry rock (HDR), from which heat would be extracted by injecting water or other fluids. A research team at Los Alamos Scientific Laboratory estimates recoverable HDR resources could rival US coal reserves.

Geothermal energy can be used for heating buildings and some industrial processes, as well as for generating electricity. Low temperature resources suitable for space heating are relatively widely distributed.

There are some environmental questions. These include land clearance in scenic areas and possible effects of massive pumping of water on adjacent wells and land stability. Indeed, proposed geothermal development in the Yellowstone region may be curtailed to protect Old Faithful.

Although geothermal power makes only a small contribution today, it is growing at a 16 percent annual rate. That compares with a mere 7 percent up to 1975. "Something dramatic has happened . . . ," says L.J.P. Muffler of the USGS. Geothermal power is taking off.

Solar

Solar energy comes not only from direct sunlight, but also from the movement of winds, the burning of wood, and the fall of water over a dam. This basic diversity is compounded by the number of different technologies that have been built or proposed to harness each energy flow.

These range from something as simple as a properly placed window to the sophistication of a solar cell designed with some of the deepest insights of modern physics.

Estimates of the percentage of the nation's energy demand that solar technologies might supply by the year 2000 run the gamut from 4 to 40 percent. Broadly defined solar now accounts for 6 percent of the US supply, and there is a widespread feeling that 20 percent (the equivalent of 7.6 million barrels of oil per day) is an achievable goal.

Of solar's current contribution, two-thirds comes from hydroelectric generation. A recent study by the US Army Corps of Engineers finds this could be doubled by putting generators on 3 percent of the existing, small dams in the nation.

As far as using sunlight directly, solar hot water heaters are considered economic in most parts of the country. Over 100,000 homes and buildings have been equipped with solar heaters: Between 25 and 50 percent of them are in California. Space heating appears feasible in some areas as well. The solar-heating industry is growing rapidly but so far remains an "elitist" market, the predominant buyers being in upper-income brackets. Use of concentrating collectors to produce steam for industrial processes is gaining momentum as well.

Wind power remains at the ragged edge of economics: It is competitive only in remote areas and has reliability problems. At the same time, the Department of Energy is continuing its program to develop large wind turbines. The largest, with a 200-foot blade span, was started this year in North Carolina. With winds at 25 m.p.h., it can generate 25 megawatts, enough electricity to power 5,000 homes.

If current promise holds true, the most revolutionary solar technology will be that of solar cells. These space-age devices, which convert sunlight directly into electricity, started at an astronomical cost, but costs are rapidly dropping. According to experts at the Jet Propulsion Laboratory, the lead center in federal efforts to reduce solar cell costs, the current price is running about $5 per peak watt. (A comparable cost for electricity from a utility is 30 to 60 cents per peak watt.)

There is a general feeling that thin film technology will be required to break the $1 per peak watt barrier. Thin film technology involves spraying a layer of silicon or a similar substance a few thousandths of an inch thick on a backing of glass, steel, or plastic. This lends itself to mass production and so to considerable cost reduction. With the 7 percent conversion efficiencies now in sight, production of 15-cent per peak watt power is a definite possibility.

The Department of Energy projects that solar cells can be generating an amount of electricity equivalent to that produced by 1.6 million barrels of oil per day by the year 2000.

NEW INSIGHTS ON TOMORROW'S NUCLEAR POWER
by Floyd L. Culler
From *The Christian Science Monitor,* December 17, 1979.

The President's response to the Kemeny commission report Dec. 7, 1979, will do much to dispel the uncertainty of direction in our national energy program.

We agree with the President's statement of the need for nuclear power to relieve our long-range dependence upon foreign-oil imports. We endorse his commitment to proceed carefully with nuclear-reactor licensing without undue delay. The nuclear utilities are prepared to work within the outlines of the President's program to achieve high standards for the safety and reliability of nuclear power.

The President's message places the responsibility upon the utility industry for doing its part in this national effort to revalidate nuclear power. The principal institutions for coordinating and implementing the utilities' programs to improve the operation and safety of nuclear plants are the already established Nuclear Safety Analysis Center (NSAC) and the Institute for Nuclear Power Operations (INPO). These organizations, along with the individual companies and their suppliers, will work with the Nuclear Regulatory Commission (NRC) as part of the team to achieve the President's objectives.

The original conception of the benefits of nuclear power now are being realized, although this important fact is obscured at times by persistent arguments that the risks of nuclear power are too great. Large numbers of nuclear electricity generating systems have been operating safely for many years in almost every major nation. Other than hydropower, the cheapest electricity is generated by these nuclear stations, and electricity generated from nuclear power saves about the equivalent of 30,000 barrels of oil per day for each

© 1979 The Christian Science Publishing Society. All rights reserved.

large plant (1 million kilowatt capacity). But the public perception of the safety of nuclear power and confidence in the institutions which manage and regulate it have been shaken by the Three Mile Island accident.

Because of the sobering event at TMI, a new cycle of evaluation of risks — both quantifiable and imagined — is under way in the United States. This re-evaluation is based on the Kemeny report, on industry studies and evaluations of great intensity, and on a major review of nuclear licensing and regulation.

The consequences of this dramatic event in nuclear power are being worked out while a rapidly deteriorating situation involving the world's principal oil resources make action to guarantee energy supplies more urgent. The serious turmoil in the Middle East now threatens about 50 percent of our oil supply, and significantly larger fractions of the supply of other major industrial nations in the world. The uncertainty which surrounds our ability to satisfy the United States's need for energy, created by these twin disasters, cries for immediate action to diminish our dependence on foreign petroleum and to quickly resolve our concerns and difficulties for using other systems. We now should act to accelerate the transition from oil and gas to other energy resources that are domestically abundant. Nuclear is available; six plants now await licensing and many others are at various stages of construction.

The very expensive transition from easy-to-burn gasoline and oil to the secondary sources of energy — coal, nuclear, solar, wind, biomass, and geothermal — must proceed without being profligately wasteful of capital. Existing systems for generating electricity must be converted or replaced under conditions which do not cause further deterioration of our economy. New capacity must be added to meet the energy requirements of an increasing population, while taking advantage of every opportunity to conserve which satisfies the simple tests of practical good sense, politically and economically. The dynamics of this change should be guided to provide energy at the lowest cost and at the lowest real risk to the environment and public health and safety.

Can nuclear power be judged to be safe enough? This must be decided in the United States during the next months, as tension mounts in the Middle East.

The President's message on the Kemeny report indicates that the national view is that nuclear power plants are safe but should be improved. He did not recommend closing down operating reactors. He did not call for a moratorium or a long de facto delay in starting those nuclear plants now ready to operate. Rather, he asked the Nuclear Regulatory Commission to proceed with licensing soon and to limit the present delay to six months. We think that the plants now being built and those new ones which will be needed, therefore, will be licensed at a steady pace as the questions raised by the Three Mile Island accident are resolved. This set of conclusions is based on our confidence that nuclear power is safe and can be operated with minimal risk to the public, particularly as revisions to regulatory processes and improvements to the plants and their operation proceed.

My reasons for confidence in the safety of nuclear power come from the established record of safety and from a look at TMI itself. But, perhaps the most important reason for confidence is the attitude of the nuclear utilities about major, increased effort to assure nuclear safety. These actions signal the coming of age of nuclear power — the maturation of a major industry. The utilities are assuming control and internal responsibility for improving reliability and safety of nuclear thechnology, rather than depending heavily upon the supplier and upon licensing processes administerd by the government.

What did we observe in the TMI incident and its followup that adds to our confidence in nuclear safety? First, from the many analyses nothing has emerged that says nuclear power is fundamentally unsafe. The TMI accident did not indicate that there is a higher probability of an accident which would expose the public to hazardous radiation in the statistical sense, but it did destroy public confidence. This public confidence must be restored by improving safety precautions technically and operationally — a process that is well started.

Despite the malfunction of equipment and the errors of interpretation and inappropriate actions by the operators, the accident was contained. Only a negligible quantity of radioactive gases was released. The Kemeny report observes ". . . We conclude that in spite of serious damage to the plant (and core) most of the radiation was contained, and the actual release will have a negligible effect on the physical health of individuals." Thus, the primary objective of all nuclear-safety design and regulation — the protection of the public — was achieved.

The accident figuratively jolted the nuclear electricity generating companies to an awareness of the full measure of their responsibility for the standards of excellence in nuclear safety. This, to me, is the most important result of TMI. It is causing a revolution, of a sort, in the utility approach to nuclear power generation and to safety. Independent utility companies are agreeing to programs of mutual support and surveillance whose purpose is to push safety to high standards of excellence. These actions include:

— The establishment of a permanent Nuclear Safety Analysis Center (NSAC) to continue technical analysis and improvement. Among other responsibilities, this center is analyzing unusual events and equipment failurs at nuclear stations as they are reported and transferring information to the reactor operators concerned. The center will continue to work on other technical improvements to safety and reliability.

— The establishment of a new institute for Nuclear Power Operations (INPO) in Atlanta will provide for comprehensive improvements in training through consistent application of rigorous criteria, tests, and retests for operators. The institute will conduct surveys of the operations at each reactor site. It will be responsible for coordinating the response to emergencies and accidents which might occur in the future.

— The provision of nuclear utility self-insurance against the costs incurred as a result of loss of power, in the event of a nuclear event. To be eligible for this insurance — which will certainly be desirable for each nuclear utility — a utility must meet the criteria and standards of excellence established by INPO, the NRC, NSAC, and others.

We are confident that, with the administration and the President backing improvement of the regulatory processes, there will be significant improvements there. Can the nuclear industry survive during the period of indecision? Yes, if delays are not too long. We believe that the many points suggested for improvement by the Kemeny commission. NRC, NSAC, and the industry can be accomplished in time to sustain our domestic capability to build new facilities. Most nuclear suppliers appear to be capable of weathering several years with few orders, but there will be losses in skilled manufacturing and design personnel. If there is a protracted delay in new nuclear plant orders, we will face a situation where our ability to build nuclear stations will be sharply reduced.

Attitudes in other countries seem to be more favorable to nuclear power than in the United States, even after TMI and the Kemeny report. With some uncertainty about my observations, I found that, in most countries, nuclear power use will proceed at the pace previously established, or will increase as the prospects for stability in the international oil supply worsen.

Unfortunately, there is a decrease in confidence in the United States as a partner or as a supplier of power systems and nuclear fuel. Because of the politics of nuclear power in the US, we are rapidly losing our well-established influence and lines of trade. This, unfortunately, increases the problems of our domestic nuclear reactor suppliers, who depend upon foreign as well as domestic orders.

Thus, I think that the electric utilities in the United States have reaffirmecd their conviction that nuclear power is essential. Further, they are saying that safety can be improved to the extent that the public risk will be further reduced. There is a certainty that, with improvements to be achieved after reexamination of technology and institutions which produce and regulate nuclear power, new levels of safety will be attained.

Dr. Floyd L. Culler is president of the Electric Power Research Institute and former deputy director of Oak Ridge National Laboratory.

WHAT ONE COUNTY IS DOING TO EARN
ENERGY FREEDOM / *by Mark Cherniak*
From *The Christian Science Monitor,* December 17, 1979.

When oil becomes too scarce to burn, when we understand that using too much coal is incompatible with human health and environment, and when all the nuclear plants in the United States are shut down, many communities in this country will hardly notice the difference. These communities are the ones

© 1979. The Christian Science Publishing Society. All rights reserved.

that are taking steps now to determine what renewable energy systems can serve the needs of their citizens.

Imagine the street you live in. Almost everyone is doing something to use energy more efficiently — whether it be adding insulation, installing storm windows, or reducing lighting and appliance use. Neighbors begin meeting to discover ways they can meet their reduced energy needs with various solar technologies. Local government lends it support by first tightening its own house and then by developing energy-efficient building and zoning codes. Bonds are issued to provide low-interest energy loans and direct assistance is given to those who cannot afford the up-front costs. A local solar industry is created to begin the massive job of retrofitting existing buildings with solar hot water and space heating systems. Solar greenhouse workshops run rampant through the town. Arrays of wind generators are planned, hydroelectric facilities are prepared, and those who can, burn wood.

Further imagine this general activity happening house by house, building by building, block by block, neighborhood by neighborhood, town by town, county by county. Then there is something happening that no one can change or control or even at times understand.

What is extraordinary about this scenario is that it is beginning to happen in many places in actual fact. And it seems that this country's transition to renewable energy resources must happen in this way if it is to happen at all.

Current federal policy with its overwhelming commitment to fossil and nuclear technologies is forcing local communities to look closely at their own potentials for surviving the energy crisis.

Rural Franklin County in western Massachusetts is one such community. It is on the path of planning and developing its local, renewable energy resources.

Twenty-three of the 26 towns in the county have established town energy conservation committees over the past 2½ years. These groups, in addition to educating townspeople, or doing projects such as weather-stripping the town library, meet monthly as the Franklin County Energy Conservation Task Force. Along with representatives of the town committees, this body includes the local gas and electric utilities, the League of Women Voters, and Cooperative Extension Service, and others.

The task force has produced an energy profile of the county, a comprehensive survey of existing, abandoned, and potential hydroelectric sites, an energy attitude survey, and, most recently, the county planning board and county commissioners have approved a five-year energy plan. It includes goals for reduced energy consumption (electricity, 3 percent; kerosene, heating oil, 34 percent; gasoline, 13 percent) and increased production of energy from renewable sources (107 percent for wood plus encouragement for solar energy, and electricity generation from water power and renewable fuels). It does anticipate a 20 percent increase in use of natural gas as a substitute for heating oil. The plan also calls for a moratorium on construction of nuclear facilities in the county — a matter that now is a live political issue. And it provides for a study of the feasibility of establishing a county energy authority. In July, 1978, with the blessings of the task force, a study of the county's energy history, conservation options, and renewable resource potential was begun by a research team at the University of Massachusetts at Amherst. Two months later a public meeting was held to announce the preliminary results.

There 150 residents were told that the county (population 63,000), with nearly one-third of the households considered low income, had an energy economy of nearly $50 million in 1975. The residential sector, including households and private automobile use, was spending $24 million in 1978. Furthermore, they were told that if things continue with only moderate energy consumption and price increases, the county would have a $225 million energy bill by the year 2000.

Just to keep even, hundreds of new jobs would be required to offset the growing number of dollars leaving the county to pay for oil, natural gas, and nuclear energy. Residents were astounded.

Last April, the energy study was completed. Conservation, it concluded, could help the county reduce its overall energy consumption by 60 percent. Renewable energy sources could meet all of the remaining demand. The biggest problem comes from transportation with the need for millions of gallons of liquid fuel. By pushing the numbers, it was determined that the county's cars could be fueled by ethanol produced from sugar beets. But it would take up to 88 percent of all the agricultural land available and the county needs the acreage for increased food production.

It was also discovered that current hydro-electric capacity in existing utility-owned dams exceeds the county's demands on a yearly basis.

Overall, the study proved valuable as a yardstick to measure progress toward greater reliance on alternate energy systems. The study also has significance as a pioneering effort in community energy planning. It also demonstrated that energy statistics *must* be gathered as locally as possible if they are to be of value in policy making.

The current emphasis on aggregate data in policy planning at the national, regional, and even state level is not adequate to determine what must be done to develop renewable sources. One cannot determine, no matter how sophisticated the computer program, the available energy at a particular site unless one is there. The literal nature of the concept of site-specific application has profound impact on how and where we develop the renewables. A particular building, stream, windy hillside, or acre of woods exists in town or city and is owned by someone. The decision to develop an energy resource can only be properly made at this most local level.

Beyond individual decisionmaking lies the new area of community energy systems planning and implementation. Across the country, communities are looking hard at municipal ownership as a means to gain control of their own energy futures. The ensuing political struggles will set the stage for years to come, until the eventual recognition that finally responsibility and accountability for community-scale systems must reside within the framework which a particular community chooses for itself. Whether it is a neighborhood, town, or county, that community is in the best position to evaluate the strengths and weaknesses of its energy activities.

WHATEVER HAPPENED TO THE IDEA OF PROGRESS?
by Richard Eder
From *The New York Times*, December 30, 1979.

A bit less than 50 years ago, about when the Great Depression began to bite, the historian Charles Beard summed up the scientific, social and technological achievements of the United States in a book called "A Century of Progress." In his preface, Mr. Beard dealt with the idea of progress in such terms as: "Everywhere it makes its way, dissolving the feudal institutions of Europe, disturbing the slumbers of the Orient, arousing lethargic Russia and finding a naked avowal in the United States of America: The earth may be subdued to the security, welfare and delight of them that dwell therein." It is not the notion of Japan sleeping that makes Professor Beard's paragraph seem so quaint, nor that of a Soviet Union in need of arousing. History naturally ages, but it is the unalloyed confidence in progress that marks the distance between the threshold of the 1930's and that of the 1980's.

The times are full of questions: Can the United States find sources for the energy it has always needed? Can it sustain a political position in a world that seems increasingly hostile? Has its productive vitality, as one writer put it, passed its climacteric and begun a long–term decline? Are bad times upon us, conceivably the worst we have known? And beyond this, how might the country's political, social and psychological fabric be expected to hold up, if the answer turns out to be yes? Fundamental to such a question is: What will happen to the long-held notion of progress in the American character? In the early 19th century, Alexis de Tocqueville remarked on the American belief in "indefinite perfectability." When he asked a sailor why American ships were not better made, "he answered offhand that the art of navigation was making such quick progress that even the best of boats would be almost useless if it lasted more than a few years" — one of the earlier affirmations of planned obsolescence. But is the American notion of progress likely to survive the next decade and, if seriously undermined, what are the likely consequences?

Responding to these questions, sociologists and social historians almost all agreed that it would be severely tested. "Progress has been he American belief," said David Riesman, the Harvard scholar. "There are paradoxes— even the Calvinists had a strain of fatalism buried in their moneymaking. But even there you had the notion of the City on the Hill and the American Commonwealth, a faith that manifested itself in activism and energy." Of course, there have been bad periods in our history that put the belief to a considerable strain. Joseph Featherstone, a social historian also at Harvard, recalled the feeling of disarray in the 1850's, when the North and South were becoming bitterly estranged: "You had a widespread feeling of despair, a general intellectual questioning of our values. Both Hawthorne and Melville felt that our ideas of progress were an illusion." Mr. Riesman recalled living in Federal camps for unemployed transients in the early 1930's. "I found a faith in prog-

© 1979/80 by The New York Times Company. Reprinted by permission.

ress that was not at all lost among these people. I realized that my Communist friends like Paul Sweezey were out of their minds to think the country was ripe for revolution. The people I met had what may have been quite a fallacious hope that things would get better — they only did get better when the war started — but they *had* it.''

"As late as the 40's or 50's'' said Oscar Handlin, the historian, ''people didn't assume that poverty or racism would go away by themselves, but they did assume it was possible to do something about them. In the last couple of decades, the feeling's grown up that there isn't very much that individuals can do, yet things will get better anyway. The idea of porgress can be a kind of narcotic, or it can be a sense of possibility which means you've got to struggle.''

Social and institutional changes since the Depression may make it more difficult for some Americans to cope with bad times. Mr. Riesman spoke of a decline of authority. It might be hard to get young people into the kind of work projects used by the New Deal. The fragmentation of political life into single-issue constituencies, and what he calls ''the growing litigiousness'' of the American temper could obstruct drastic measures for re-ordering resources. ''In the Depression you still had the capacity for national management of the crisis,'' said Daniel Bell, the sociologist. ''Today, problems are embedded in an international economy; monetary policy is restricted by billions of Eurodollars sloshing around; there's the Third World problem, the problem of commodity cartels, not merely in oil. Our national state is too big for small problems, and too small for big ones.''

Narcissism Can Be Helpful

Less gloomy, Philip Converse, a University of Michigan sociologist and historian, suggested that however bad things get, some aspects of our technology, particularly for information and communications, should continue to grow. But he agreed that a considerable ''jolt,'' material and psychological, was almost inevitable. How would Americans react if the political stresses of the past decade were to be capped by a harsh economic battering in the next? ''One of the difficulties,'' said Mr. Bell, ''is that when you substitute an ethos of material growth for more traditional values, when you face a setback, it's harder for people to find something to fall back — the will of God and so on. It tends to break down the sense of *civitas,* to create an attitude of devil-take-the-hindmost.''

Michael Maccoby, the psychiatrist and sociologist author of ''The Gamesmen,'' sees persistent signs of hope. The self-examination many Americans have gone through in the past decade, he says, may help them to weather bad times. The emphasis on health and spiritual development, sometimes criticized as narcissism, may in fact be useful preparation. ''We've lost the right to the unlimited use of our cars, to eating steak four times a week. But it's easier to give up steak because of cholesterol. It's easier to reduce the use of the car in the interests of health to walk, to bicycle. Narcissism, up to a point, can make us more adaptable.''

Dr. Maccoby stressed the need for leadership guided by ethical values. A sense of justice is crucial when people must make sacrifices. "Ten years ago I interviewed a number of leaders. When I asked them what their principles were, they said, 'What do you mean? What we need is flexibility.' The problems we will face in hard times will be of setting limits. We cannot live without limits, and intelligent people realize this; but we must define these limits before they are thrust upon us. And to do so requires a sense of ethics, the leaders who can act out of principle, rather than organizing adversarially to defend some particular set of interests."

Dr. Maccoby's talk of the need for a leadership of principle to hold together a country that may have to relinquish, for a time, the hope of a leadership of success, recalls part of an article by Mr. Bell: "America was the exemplary once-born nation, the land of sky-blue optimism in which the traditional ills of civilization were, as Emerson once said, merely the measles and whooping cough of growing up. The act of becoming twice-born, the entrance into maturity, is the recognition of the mortality of countries within the time scales of history."

CHURCHES FEELING ENERGY STRAINS IN THEIR BUDGETS AND THEOLOGY / *by Kenneth A. Briggs*
From *The New York Times,* January 27, 1980.
Special to The New York Times

Many religious congregations, their budgets already strained, are finding that rising fuel costs are changing their worship patterns and programs and, in some cases, helping determine whether they survive.

Churches in inner cities and rural areas have been hit especially hard. "The institutions are hurting," said the Rev. Bernard A. Holliday, the energy specialist at the Council of Churches of the City of New York. "If many of them are not careful, they will go under." Others simply do not have as much money to use for charitable purposes, Mr. Holliday said.

The situation has also raised ethical and theological questions about the churches' role in helping Americans justify material expectations in the past and helping them induce a spirit of sacrifice in the future. A Roman Catholic bishop told a recent interfaith conference that energy shortages would create the "the pre-eminent social justice issue of the 1980's."

Gradually, with architectural guidance from denominational headquarters and with such books as "The Energy Efficient Church" by Douglas Hoffman, churches and synagogues around the country are learning to caulk and insulate, to turn down the thermostats and to hold services in small chapels, rather than vaulted sanctuaries, whenever possible.

In addition, some churches have begun to share space with other congrega-

© 1979/80 by The New York Times Company. Reprinted by permission.

tions, and others have sought to fight the rate increases that could hasten financial collapse.

And some have taken the situation as an opportunity for innovation and renewal. For example, the First Presbyterian Church on Staten Island initiated its fuel-saving plan with a special observance that described goals and playfully featured young people modeling long woolen underwear.

Drain Can Be Enormous

But church buildings swallow up prodigious amounts of heat, and, even with these efforts, the fuel-related drain on the budget can be enormous.

The Old Orthodox Church of the Nativity in Erie, Pa., for example, installed thick thermal-pane outer doors, turned the thermostat down to 64 degrees and replaced traditional leaded stained glass with faceted-glass windows lined with foam rubber. Nevertheless, gas bills have increased by 400 percent since 1976, to $4,500 from $912. The church has managed to keep up with costs, but many others walk a thinner line.

At St. Charles Lwanga Church, a black Roman Catholic parish on Chicago's South Side, fuel costs have damaged programs to feed and clothe poor people. The pastor, the Rev. Paul Burke, has appealed to suburban parishes for help in maintaining the programs. The parish now runs bingo two nights a week rather than one to help meet added costs.

In Detroit, St. Patrick's Roman Catholic Church heats its sanctuary only on Sundays. The gas company's recent request for a 22 percent rate increase, says the Rev. Thomas J. Duffey, would force the closing of the church and its allied center for the elderly.

Kingsley United Methodist Church in Milwaukee is one of the small number of congregations in the nation that have decided to disband, citing rising fuel costs as one of the leading factors. Members of the half-block-long church in the central city are mostly elderly, and membership has declined in recent years. With financial problems getting worse, the congregation voted to disband as of June 15.

While the need to conserve energy is by far the most immediate concern of the religious groups, their discussion of the issue gets into moral and theological dimensions. Religious thinkers are asking how spiritual resources can help reshape American values into an "ethics of scarcity" that emphasizes simpler, less material ways of living. Many religious leaders are also drawing attention to the additional burdens that higher energy rates are imposing on the poor.

"Value questions are being taken seriously and they aren't being lost in the practical worries," said Chris Cowap, the staff coordinator of energy policy for the National Council of Churches. "When I go out to speak, I find that what people want to hear is what the ethical and theological rationale are for the church's involvement in the issue."

At a national conference on "Religion and Energy in the '80s" in Washington this month, Bishop William M. Cosgrave of the Roman Catholic Diocese of Belleville, Ill., said:

"I submit to you that the most important thing religious institutions can do is to communicate some of the human reality of the energy situation and motivate a human response to it. We are not talking about abstractions and statistics — or we shouldn't be. We are talking about war and famine and suffering. We are talking about the struggle against cold, against dark, against isolation.''

Carter Joins in the Plea

President Carter and several other speakers urged the 40 organizations at the interfaith conference to build a spiritual basis for a reduced standard of material life.

"It might seem strange to some, not to you, that the conservation of oil has a religious connotation,'' Mr. Carter said, "but when God created the earth and gave human beings dominion over it, it was with the understanding on the part of us, then and down through the generations, that we are indeed stewards under God's guidance.''

Dr. Elizabeth Bettenhausen, a professor of social ethics at the Boston University School of Theology, expressed a growing theme in religious circles when she told the conference that religion must change some basic assumptions that churches had helped instill in earlier ages. In previous times in America, Dr. Bettenhausen said, churches generally taught a theology that said "the physical world is distinctly inferior to the world that awaits us in the by-and-by.''

At the same time, Dr. Bettenhausen said, it was common for Americans to foster a "reward system theology'' that assumed "God loves America in a peculiar way and that our problems were only temporary.'' She added, "How do we now motivate people to cut back when their theology has promised them unlimited abundance?''

Rabbi Marc H. Tanenbaum of the American Jewish Committee believes that the concept of Americans as the "Chosen People'' has made the idea of sacrifice distasteful. "Beneath the biblical metaphor was laid the foundation of a pervasive ethic of materialism,'' Rabbi Tanenbaum said. "But that ethic today has become a source of contradiction.''

A NATIONAL PREOCCUPATION

From *U.S. News—A World Report,* December 31, 1979/January 7, 1980.

For the third time in a decade, Americans are being called upon to conserve their way out of an energy crisis.

President Carter is asking motorists to cut down on driving and obey the 55-mile-per-hour speed limit. He has ordered thermostats lowered this winter in most buildings frequented by the public and has urged Americans to follow suit in their homes to offset the loss of Iranian oil, restrain inflation and reduce pressure on the dollar in the world market.

The new drive comes at a time when U.S. businesses and homeowners, already convinced by the 1973–74 Arab oil embargo that an era of cheap and plentiful energy is at an end, are taking impressive steps of their own to conserve energy.

American-made autos now get 53 percent more miles per gallon of fuel than they did in 1974. Millions of homeowners have insulated their houses and installed storm windows and other energy-saving devices. Thermostats have been dialed down to 65 degrees this winter in homes and buildings throughout the country. Airlines used only 3.8 percent more jet fuel in 1978 than in 1973 while carrying 100 million more passengers.

Car-pooling is on the increase among urban commuters. More police officers are walking their beats. Many mail carriers are delivering mail on foot and bicycle instead of by fuel-powered vehicle.

Nationwide, oil use is running 4.9 percent less than a year ago, including a 9.7 percent drop in gasoline demand. The gross national product—the total value of goods and services—increased over the last five years at a rate of 4.4 percent annually, while energy consumption grew at a pace of only about 2 percent a year.

What it all means, says Energy Secretary Charles Duncan, is that the U.S. has "broken the historic link between economic growth and growth in energy consumption." Following is a close-up look at exactly how businesses and individuals are saving energy.

Industry: The Little Things Pay Off

American industry has been searching aggressively for several years for ways to cut fuel use—with major results.

Energy used by industry has increased less than 1 percent a year since 1972—the last full year before the Arab embargo. In the same period, industrial output increased 21.1 percent and industrial employment grew by 1.7 million workers. Industry's share of total energy consumption has fallen from 38.6 percent in 1972 to 35.9 percent in 1978, and the trend is continuing.

General Motors Corporation since 1972 has reduced the energy cost of producing a car by almost 1½ barrels of oil.

American Telephone & Telegraph Company now handles 24,300 phone calls with the energy equivalent of a barrel of oil. Five years ago, that barrel produced 17,700 calls.

Before the 1973 embargo, Exxon Corporation refineries consumed 1 barrel of oil of every 10 barrels processed. Today, that figure has been reduced to 1 of 13 barrels.

Some industries have exceeded even the stringent goals set by the Energy Policy and Conservation Act of 1975. The law gave manufacturers of transportation equipment until 1985 to cut energy consumption by 16 percent from the amount used in 1972—a goal that was surpassed in mid-1977. Nonelectrical-machinery producers were targeted for 15 percent energy savings by 1985, and achieved a cutback of 21.3 percent by 1978.

AT&T executive W.M. Ellinghaus explains why industry is so interested in saving energy: "In 1973, the Bell System spent 275 million dollars on energy. Last year, the figure was 578 million dollars. And we estimate our total energy costs for this year at about 675 million dollars, with no increase in energy consumption. Our projection for 1980, even though we intend to hold total energy use at its present level, is in the neighborhood of 810 million dollars."

Reports Berndt K. Lyckberg, corporate manager of energy for Firestone Tire & Rubber Company: "Rising prices have tripled Firestone's total energy costs since 1972, making conservation measures vitally important."

Much of the energy saved by industry results from simple "housekeeping" chores—turning off lights, checking thermostats, closing windows and performing routine maintenance on cooling and heating systems. Examples abound:

—Mamma Leone's restaurant in New York reduced its steam-heat requirements by 27 percent after an energy audit discovered a leaking steam trap.

—The J.I. Case Company of Racine, Wis., manufacturer of farm tractors and construction equipment, is saving $46,000 a year in energy costs after switching from hot to cold water to wash parts.

—The owners of a 21-story Chicago office building spent $213,000 to clean and repair heating and cooling equipment. Utility bills for the building were reduced so rapidly that the owners recovered their investment in three years.

—AT&T reports a substantial reduction in energy use by its vehicle fleet simply by keeping car and truck engines tuned and by buying subcompact cars with standard transmissions. The Bell System's 186,000 vehicles are burning only 1.3 percent more fuel than in 1973, although the number of vehicles has risen 12.3 percent.

—Tests sponsored by the trucking industry found that trucks can cut fuel consumption by 32.2 percent if drivers obey the 55-mph speed limit. Continental Baking Company believes it can save even more by installing controls that will limit truck speeds to 45 mph. The company also is switching to tubeless tires and is replacing double rear wheels with single wheels to reduce road friction.

—The Westinghouse Research and Development Center in Pittsburgh is saving 42 million cubic feet of natural gas and 7 million kilowatt hours of electricity a year by turning off lights and lowering lighting levels, reducing water temperatures, plugging air leaks and turning off ventilation during nonworking hours.

—Mutual of Omaha Insurance Company has cut its fuel consumption by 50 percent, mainly by turning off its heating system and relying on heat produced by sun shining on the windows and the body heat of the firm's 4,600 employes. "In fact," says John Cleeton, chief mechanical engineer, "it gets so warm sometimes we have to cool things down by bringing in outside air."

By taking such steps, says E. Milton Bevington, president of Servidyne, an Atlanta energy-management firm, "most companies can cut heating and cooling costs in buildings by 40 percent and recover their investment in two years."

From Sunlight to Computers

Other companies have realized big energy savings from rather small investments of capital.

The Northrop Corporation, for example, installed 250 skylights at one of its Los Angeles plants, making it possible to turn off some of the lights between 10 a.m. and 2 p.m. each day. Result: A 10 percent energy saving.

The Marriott Corporation has cut its utility bills by 15 to 38 percent in 75 restaurants. One method: Installing special dishwashers that operate effectively at low temperatures, require less water and use chemicals instead of high water temperatures for cleaning.

Many firms have discovered that computers can reduce energy use. Dillard Department Stores has slashed utility bills by 20 percent after installing a computer to monitor and control building temperatures. At Dillard's home store in Little Rock, the saving amounts to $414,000 a year.

Trans World Airlines plans to install in its aircraft computerized systems that will make minor adjustments to help planes fly more efficiently. TWA expects the devices to reduce fuel use by 4 percent—a sizable chunk, given the billion dollars the airline expects to pay for fuel in 1980.

Monsanto Company is spending an average of 50 million dollars a year on energy-conservation projects. So far, the outlays have resulted in a 20 percent reduction in energy use at Monsanto's chemical plants since 1972. The target for 1985 is a further 15 percent cut in fuel consumption.

Business executives warn that future energy savings will come harder and require investment of billions of dollars in new plants and equipment.

"Routine housekeeping measures to save energy already have been taken," says W. William Pritsky, technical director for the Aluminum Association, which recently set a goal of reducing energy consumption by 20 percent through 1985. "Our goals for additional reductions will require large-scale capital expenditures."

People: Common Sense and Ingenuity

When their electricity bills rose to $140 a month last winter, Robert Hawkins and his wife decided it was time to install a more cost-efficient heating system in their 30-year-old home in Sacramento, Calif.

They replaced their old-fashioned electric wall heaters with a $9,000 solar-heating-and-cooling system that Hawkins, a self-employed research consultant, installed himself. The cost was considerably more than the $5,800 expense of a conventional heating system, but the investment is paying off rapidly with a 55 percent solar-energy tax credit and a 75 percent reduction in the electricity bill.

The Hawkinses' experience is fairly typical of millions of American families who are taking steps—many of them highly ingenious—to hold down soaring energy costs this winter. Projects run the gamut from caulking windows and adding wood stoves to trading gas-guzzling cars for more-efficient subcompacts—even redesigning entire houses.

Many people are planning vacations involving less travel by private car, a trend reflected in a drop in requests to the American Automobile Association for trip routings. Home-repair centers, especially in the Northeast, report brisk sales on insulation, storm doors and weatherstripping. Wood-burning stoves are a hot fad across the snow belt.

Sales of mopeds—motorized bicycles that get better than 100 miles to the gallon—are soaring. A 40 percent increase is reported in demand in the New York area for Amtrak rail travel. Car pools are returning to levels reached during the 1973–74 Arab oil embargo. Nearly 10 percent of all home-owners have added insulation in recent months.

In areas where heating fuel is especially costly, group efforts are under way to cut consumption. Fitchburg, Mass., a town of 38,000 northwest of Boston, is experimenting with a program to reduce home-heating bills by 25 to 30 percent through window caulking, weatherstripping and insulation provided at cost to most families and free to those with incomes of less than $14,000. The plan was initiated as a pilot project by the federal volunteer agency ACTION.

Householders in the Midwest are snapping up all sorts of energy-efficient devices to help heat their homes. Says Bob Abele, president of the Energy Store in Northbrook, Ill.: "Business is fantastic now. People are sick and tired of the utilities, and the gasoline crunch has made people more aware of the need to conserve." One new product that has "really taken off," Abele reports, is the "window quilt," a 99 percent opaque window covering that is similar to a shade but which seals more tightly and insulates a window.

"It's going to be a good year for the solar market," predicts Terry Turzynski, president of Natural Energy Systems in Palatine, Ill. "Solar products used to attract people with a lot of money, but now they're starting to move into middle-income brackets. People building $70,000-to-$80,000 homes are starting to look at solar."

Turzynski adds: "There's been a tremendous increase in sales of a fuel-efficient fireplace called a water-jacketed firebox." Water heated by the firebox is carried to a heat exchange in the home furnace to help heat the house.

A Houston secretary tells of spending four weekends and a week of her vacation caulking windows, sealing doors, filling in cracks and holes around pipes and insulating electrical switches and plates. "If I can do it, anybody can," she declares. "Once you become sensitive to the feel of your house, it can really make a difference and you know it at once."

Back to the Good Old Days

In Hanson, Ky., Mack Tyner, an intern at the Trover Clinic in nearby Madisonville, and his wife Paula have installed two wood-burning stoves that heat their entire house. They adapted a copper-coil heating system to one stove for heating water in the winter, adding a solar water heater for use in summer. Then they installed triple-glazed storm windows to reduce heat loss. Now he's toying with the idea of building a wind-powered generator to supply electricity. The Tyners also bought a diesel-powered subcompact car to cut the fuel cost of driving 20 miles to work.

Like many other retired couples, the Bruce MacInnises of San Francisco laid out a careful plan to trim their energy costs this winter. They proposed to first insulate the attic of their two-story home. "It'll cost $400, but it will also save us about $80 a year," MacInnis notes. Then they planned to put weather-stripping around doors and windows. But they decided against insulating the walls, because it would take about 25 years to recoup the investment. The couple rarely drive their two cars these days, using public transportation instead to hold down gasoline consumption.

Pooling rides is an idea that many find pays off handsomely. Don Torluemke figures he saves at least $2,000 a year by van-pooling from his home in Alamo, Calif., to his job at Lawrence Livermore Laboratory 25 miles away. He now owns a 15-passenger model purchased last June and charges his riders $35 a month to cover gasoline and maintenance. Torluemke thus rides free and has the van for his personal use on evenings and weekends. He receives a tax credit for interest on the purchase loan.

A cabdriver in Hollywood, Calif., Tom Bush, reports that he and his wife now do most of their personal travel on a motorcycle, which averages 50 miles to the gallon. "We even go grocery shopping together on the motorcycle," Mrs. Bush says. "But two bags is the maximum."

Riding the bus can be rewarding monetarily, too. Robert Rieke, a bank officer in Houston, figures that he saves $1,500 a year in commuting costs.

The techniques for holding down home-heating expenses seem almost endless. Robert Troy, a 38-year-old Atlantan, is using what he calls "older energy alternatives"—burning a combination of wood and coal in his fireplace to cut his gas bill by 20 percent. "Not too many people think of burning coal in a fireplace," Troy says. "But coal burns all night and it's a good supplement. Besides, $25 worth of coal will carry me all winter."

Other means are being used by the Maurice Rudiselles of Morrisville, Pa. After putting in a solar hot-water system, they now plan to convert a screened-in porch into a "passive solar design" extra room. They also have installed an automatic preset thermostat, timed to shut off when the family is not home, and an automatic flue that shuts down when the fireplace is not in use.

Charles Langland, a systems analyst in Houston, wrestled with cost-vs-payback figures a long time before installing $3,300 worth of storm windows in his home recently. Now he finds that his heating system rarely switches on, despite near-freezing temperatures. Langland has taken other steps, too—wrapping his water heater, insulating wall plugs and switches, caulking windows, installing door seals and adding a glass fireplace screen.

Another couple, Jesse and Helen Reed of Roswell, N.M., recently completed installing storm windows and insulated draperies throughout their house, then went a step further by planting ivy for more insulation on the north side and adding cypress trees as a windbreak.

An experimental device that makes use of the heat from a clothes-dryer exhaust to help heat the house is being tried by Diane Steel and 50 other householders in Xenia, Ohio.

The Richard Wicks of Roseland, N.J., have installed a thermostat in each room of their house, enabling them to shut off the heat in rooms not in use. They also keep the general house temperature around 60 to 65 degrees. They shut off their freezer about a month ago when they found it did not warrant the energy cost.

Redesigning the American Home

The prize for energy conservation in the home may go to Robert Holdridge, a 35-year-old architect of Hinesburg, Vt., who designed his family's four-bedroom, two-story house as a passive-design solar structure. Its principal source of heating energy is the sun, which provides an estimated $1,500 yearly saving in fuel costs.

The home has 750 square feet of glass paneling on the south side to take advantage of solar heat and no windows at all on the north side. The structure is fully insulated, with concrete floors and brick interior walls to retain heat. Holdridge has added a wood-burning furnace for additional heat and a wood-burning stove for cooking. Appliances that require electricity are used only during off-peak hours when electric rates are lower.

In the South, wood is replacing oil in a big way. Take the case of Janice and Bill McKinnon's renovated, two-story 1889 Victorian house in southwest Georgia. It has 10 rooms and four baths and is heated solely by fireplaces and five Norwegian wood-burning stoves. "We don't even have an electric heater," Mrs. McKinnon says. Neither do they have any air-conditioning units; an attic fan and ceiling fans cool the house effectively during the summer. The young couple's total heating bill this winter is estimated at $125 for five cords of wood; McKinnon chopped another cord himself. That compares with $100 a month they spent to heat a previous house by conventional methods.

As more and more Americans take steps to cope with the rising cost of energy, there are signs of long-range effects on house designs. In San Diego County, a trend-setting regulation requires all future homes to provide for solar water heating. A study by the University of Michigan predicts a trend toward fewer windows, shifting of major glass areas to face southern sun exposures and new emphasis on high-rise apartment houses for fuel savings of at least 15 percent.

More changes forecast: A return of such energy-conserving devices as awnings, double doors in vestibules and double-glazed windows. Home-building standards, meanwhile, have been tightened to conserve energy in such areas as California, Minnesota and Ohio.

Conservation moves in 1979 cut the annual growth rate in home electrical use in half, from 6 percent to 3 percent.

Looking ahead, Department of Energy officials predict that energy-saving steps, plus tighter government standards for home building, will reduce the energy used in American homes by 50 percent by the year 2000—marking a massive change in the country's way of life.

Copyright 1979 U.S. News & World Report, Inc.

ENERGY AND THE POOR
by Bob Swierczek and David Tyler
From *Christianity and Crisis,* Vol. 38, No. 15, October 16, 1978.

Perhaps no other domestic group has been as seriously affected by the energy crisis as the poor. Whereas in 1976 expenditures on home fuels constituted from 15 to 25 percent of an average low-income family's budget, data recently submitted to the Community Services Administration suggest that today the proportion may be considerably higher. In some parts of the country, fuel expenditures now constitute from 30 to 50 percent of a low-income family's budget. Poor families in such areas pay more for energy than for any other essential—food, rent or clothing.

In contrast, middle-income families pay on the average only from 4 to 7 percent of their income for energy. Moreover, unlike the middle class, the poor have little or no discretionary income with which to absorb energy cost increases; nor can they respond to these increases by reducing energy consumption. On the whole, poor people do not own appliances like dishwashers or washing machines whose more conservative use might reduce energy costs. In 1975 the average US low-income household used 55.4 percent less electricity and 24.1 percent less natural gas than the average middle-income household.

For the poor, then, a continuing rise in energy prices means a constant reduction in other essentials—less food, less clothing—or a reduction in essential energy use—less heat, less light.

These harsh realities are simply not addressed by the broad energy policies of current or previous administrations. The Carter policy of encouraging price increases as a goad to energy conservation, for example, disregards the inability of the poor either to pay or to conserve. It appears that this disregard, at least on the part of the current Administration, arises from oversight rather than from any conviction that the energy problems of the poor are insoluble. The solution, nevertheless, will require sustained attention, political will and imaginative yet practical programs.

What differing impact might specific energy policies have on the poor? Limited policies that encourage voluntary energy conservation or that attempt to coerce it through pricing mechanisms are, as we have seen, grossly inequitable. Policies that bring adjustments in the energy delivery system, emphasizing marginal conservation and cost-free optimal home insulation (weatherization), are desirable and can in fact reduce costs in low-income homes by up to 50 percent. But if energy costs keep increasing, a point could well be reached at which even the most thorough weatherization will still leave unacceptably high bills to be paid by the poor.

Ultimately, then, if the situation of the poor is to be addressed, energy systems must be evaluated in terms of their ability to reduce costs, taking into account both pricing mechanisms and factors of source of supply and delivery.

This means that policies depending on centralized, capital-intensive, high-technology systems (nuclear, centralized solar or solar satellites, for example)—whose high costs are justifiable only in terms of the enormous future costs anticipated for imported or domestic oil—are not likely to alleviate the energy crisis for the poor. One costly source is simply replaced with another.

Moreover, such policies envision no change in delivery systems or pricing arrangements. The cost of long-distance power delivery can be especially high for those rural poor in remote areas. The present widespread use of "declining block rate pricing"—which encourages energy use by charging less per unit for greater use and more for less use—is also inequitable for the poor, who cannot afford to use more than they do. Finally, even if a centralized system turns to another energy source, its operating costs will continue to rise as labor and other maintenance costs increase.

For the poor, the development of policies encouraging decentralized, labor-intensive, small-scale solar energy technologies combined with optimal weatherization seems to offer the best opportunity for low-cost efficient energy systems. No great capitalization is required, maintenance is low, and the system pays for itself with energy savings in 10 to 20 years. To be successful, however, such policies must recognize the limited ability of the poor to finance the change. A program of solar development through the provision of solar grants (as opposed to low-interest loans, which are not affordable) offers the best opportunity for large-scale development of solar energy among low-income people, and one realistic approach to the low-income energy crisis.

A pricing system that would reward conservation instead of encouraging consumption would also benefit the poor. "Lifelining"—charging a very low rate for the amount of energy needed for survival—is also very much in order.

Energy and Development

If the effects of the energy crisis on the lives of 12 million domestic poor are severe, the bleak energy situation that confronts the global poor is distressingly worse. Totaling about 1.2 billion and concentrated mostly in South Asian, West African and sub-Saharan countries, these peoples are victims of a vicious circle of chronic hunger, malnutrition, sickness, high birth rates and low life-expectancy, widespread illiteracy, lack of education and training, and severe unemployment and underemployment. Even before the 1973–74 OPEC-induced petroleum price increases, the poorest developing nations were achieving only minimal economic growth. Since 1968 their average per capita incomes have increased at a 1.5 percent rate—or $2 per year.

Future prospects are not appreciably better. Since 1973 the virtual quadrupling of petroleum prices has seriously jeopardized the modest socioeconomic development goals of these countries by making the importation of needed petroleum products prohibitively costly, negatively affecting the terms of trade and intensifying the search for already scarce traditional energy resources. The supply of formerly inexpensive and locally available sources such as firewood, a basic energy source for 9 of 10 residents of the poorest countries, is now largely depleted.

The intense search for firewood has also resulted in rapid and large-scale deforestation, top soil erosion and silt deposits in rivers and at dams. Moreover, sharp cost increases for firewood have led to the substitution of animal dung for cooking fuel, thus reducing its more productive use as fertilizer and causing lower agricultural output. The very survival of these nations' rural populations has been put in question. The recent Sahelian drought/famine is a prime example.

For developing countries, energy is a means to achieve the goals of development. Development programs of the 1960's emphasized a Western-oriented model of urban industrial development, using capital-intensive, high technology projects, the benefits of which were expected to "trickle down" to the urban and rural poor. In fact, such modest successes as were achieved by this approach totally bypassed the intensely populated, chronically poor rural areas. More recently development planners and practitioners have formulated a new "basic human needs" strategy which emphasizes achieving both greater equity and healthier growth by meeting the minimal basic needs of the majority rural poor. These basic human needs include the physical requirements of adequate nutrition, clean water, sufficient shelter, clothing, sanitation and light, together with such necessities as physical and mental health, personal mobility, education, and meaningful and productive jobs. Basic human needs projects are urban-and-rural-poor oriented, low-cost, labor-intensive, small-scale, technologically appropriate to family and village conditions, culturally acceptable to their communities and ecologically sound.

Energy Options

As Denis Hayes has succinctly stated, "The Third World faces difficult energy choices. With little capital, few trained technicians, scanty infrastructure, inadequate reserves of conventional fuels, and a large and rapidly growing population, it has little margin for error. Any commitment of resources to one energy option will necessarily deny them to another."

Here is a brief summary of energy policy alternatives and potential impacts on the poor.

Fossil fuels: The use of coal, oil and natural gas is not an attractive option for the developing countries because it involves high cost and complex technologies, creates unnecessary dependency on the West and does not benefit the poor. Coal is not known to exist there, except in India. Apart from the OPEC countries, Angola, the Congo, Zaire, and, perhaps, Mexico, oil is not now plentiful. Even if vast deposits are discovered and exploited, it might be wiser for poor countries to sell this production at high prices to countries that are dependent on it, as Denis Hayes suggests, and invest the revenues in renewable energy resource development. Such a strategy could have a favorable impact on the poor.

Nuclear energy: The nuclear option appears to hold a considerable attraction for certain developing countries that seek an alternative to excessively expensive petroleum systems, who view nuclear as a prestigious advance technology, and who are interested in it for power-political reasons. China, India,

Pakistan and Taiwan now have nuclear reactors; Brazil, Iran and South Korea appear committed to their development. Requiring highly centralized operations and controls, technologically very complex, environmentally unsound and very costly, nuclear energy development is not a reasonable option for poor countries. Because nuclear systems are capital-intensive, large-scale and expensive, they are not likely to benefit the rural poor.

Renewable energy resources: The use of renewable energy resources has broad potential for the Third World because in many cases there is an abundance of sunlight, dispersed rural population and traditional reliance on the sun for warmth and drying, firewood for heating and cooking and forage for draft animals. As early as 1892 Chile developed a solar still to produce clean water, and by 1912 Egypt had produced and installed a solar water pump for irrigation.

Considerable research and development in renewable energy resources is taking place in Africa, Asia and Latin America. Solar cells have been successfully used to provide electricity in India, Niger, Chile and Upper Volta. Water power has been developed in Afghanistan and Turkey, and it is estimated that China has built approximately 50,000 small-scale units to produce hydroelectricity. Argentina, Tanzania and Zambia have successfully experimented with windmill development. The use of plants—biomass—to provide energy is well underway in Brazil, where sugar cane and cassava are being harvested to produce ethanol. In the Philippines coconut husks will be used to fuel electric power plants. Yet another alternative is biogas technology, employing anaerobic bacteria to digest animal dung and human excreta to produce methane for cooking and fertilizer for farming. China is estimated to have produced 4.3 million working boiogas units.

The development of renewable energy resources will benefit the poor to the extent that projects are small-scale, decentralized, labor-intensive, low-cost and technologically appropriate to family and village scales. It is essential, however, that developing states make firm political and economic commitments, both to the development of such energy systems and to the participation of the poor in them: As Arjun Makhijani has put it: "The problems of poverty have been difficult to solve not because the poor countries lack money, oil, resources, or whatever, though they may be aggravated by these things. They have been difficult because policy makers have not been protagonists of the poor."

Finally, how ought the United States aid the poorest countries to meet their energy objectives? American assistance should take several forms, including: increased funding of domestic renewable energy resources research and development efforts in small-scale, appropriate energy technologies in poor countries; the provision of emergency energy assistance to poor countries undergoing temporary economic/energy/ecological disasters; the provision to developing countries of individuals skilled in renewable energy resources project management, through domestic private voluntary organizations and such agencies as ACTION and AID; and the provision of training and technical assistance to organizations and agencies engaged in renewable energy resources development in poor countries, including the training and education in the US of students and technicians from those nations.

Bob Swierczek and David Tyler are research associates at Design Alternatives, Inc., a consulting firm specializing in program and policy development in energy alternatives for such clients as the National Science Foundation and the Office of Technology Assessment of the Department of Energy. The authors wish to thank Eugene Eccli, president of Design Alternatives, for editorial assistance and Amatullah Sharif for her tying.

RELIGIOUS GROUPS UNITE BEHIND BANNER OF CONSERVATION, RENEWABLE ENERGY SOURCES
by Constance Holden
From *Science,* Vol. 207, 25 January 1980.

Wasting energy is definitely contrary to God's will, but this country is sorely in need of a "new technology" that promotes the values of conservation and renewable energy sources. That was the message of an unusual meeting held in Washington on 10 January on "Religion and Energy in the '80s." Inaugurated at a White House breakfast featuring President Carter and Energy Secretary Charles Duncan, the meeting was attended by about 150 representatives of the Protestant, Catholic, Orthodox, and Jewish faiths.

No major politician, with the possible exception of California Governor Jerry Brown, has tried seriously to confront the citizenry with the notion that finite resources inevitably mean changes in life-styles. So now it seems religious groups are preparing to assume the responsibility. And while they do not have the television networks at their command, there is a not inconsiderable audience out there in the churches and synagogues.

According to a spokesman for the National Council of Churches (NCC), the main sponsor, the meeting had its genesis last July when Jimmy Carter summoned a stream of people from all walks of life up to the mountaintop at Camp David to figure out what to do about the national malaise.

This month's meeting had the blessings of Carter, who was reportedly very eloquent at the breakfast, describing how conservation might improve the quality of life by fostering a new sense of community. Duncan, too, encouraged the clerics—sounding "more outspoken on conservation than Schlesinger ever was," according to one participant.

The rest of the meeting ranged from lofty moralizing about the "theological imperatives" of energy to homely talk about weatherizing churches in Connecticut.

Although energy policy has caused furious debate among various Protestant

© 1980 American Association for the Advancement of Science.

denominations, it appeared at the meeting that all three major religious groups had finally coalesced around an energy ethic that features conservation as the number one concern. Closely following is emphasis on development of renewable resources and on choices that will enhance employment opportunities and not penalize the poor.

All that is needed now, according to the main speakers, is a new theology that promotes these objectives. "We operate with a theology that makes it difficult for us to confront" the energy problem, said Elizabeth Bettenhausen, associate professor at Boston University School of Theology. She said one of the difficulties is that many Christians still regard this earth as unimportant since it is only a temporary stop on the way to heaven. Another negative factor she cited was the idea that America is made up of God's chosen people—that "God loves America in a peculiar way . . . so God will take care of everything including the limitations on natural resources." The notion that God sets limits, she noted, contradicts 250 years in which the consciousness of Americans was molded by the limitless frontier.

Rabbi Walter Wurzburger quite agreed that we need to "create the kind of theology that makes us responsible custodians of limited resources" and suggested that Judaism, with its emphasis on living in the here and now and prohibitions on waste, offered appropriate guidance. The Catholic representative, William Millerd, S.J., added that if we are seeking energy modes that fill human needs, conservation and renewable energy are desirable because they provide more employment than capital-intensive energy sources. Harold Bennett of the Southern Baptist Convention weighed in with a call for "an ethic of parsimony." The theme predominating in these talks was that when God gave man "dominion" over the earth, He did not mean man was supposed to "dominate" and gobble everything up. He meant it was man's responsibility to exercise careful stewardship over resources.

There are probably few public officials whose outlook so closely conforms to God's as Denis Hayes. Hayes, organizer of the 1970 Earth Day and now director of the Solar Energy Research Institute (SERI) in Golden, Colorado, buoyantly expressed the hope that SERI "will become for energy policy what NASA has become for space." Pricing determines what energy choices are made but the "biases in the market are staggering," he said. Federal, state, and local governments, for example, subsidize nonrenewable energy sources to the tune of $13.5 billion a year, leaving $0.5 billion for renewable ones. Yet if the government ceased treating energy technologies as "neutral and interchangeable" and instead eyed them for nonmonetary "values" such as safety, accessibility, and benign social and environmental effects, it would see that it is backing the wrong horses.

The talk of values that always accompanies talk of soft, or alternative, or nonrenewable energy is symptomatic of the fact that ethical considerations have only recently penetrated some areas of national discussion. The civil rights and war movements awoke Americans to reassessing their moral and ethical assumptions. Many a university stock portfolio was reshuffled as a result of pressures to have investment policies conform with goals of humanitarianism and social justice. Now these righteous preoccupations have spread to energy under the heading of "ecological justice"—defined in an NCC pamphlet as "equity for all members of the community of life within the sustainable boundaries of the biosphere."

A stockholder action for ecological justice is being supported by the Interfaith Center for Corporate Responsibility, a group sponsored by the NCC. According to director Timothy Smith, 12 utility companies across the land are to be petitioned by religious groups that are also shareholders to add three items to the ballot at the annual shareholders meeting: people will vote on whether to have the company look harder for alternative (that is, renewable) energy sources, to put greater emphasis on energy conservation, and to stop all development on nuclear plants.

Energy has been a focus of the NCC's moral concern since 1976, when it issued a resolution calling for a ban on plutonium reprocessing and its use for energy production. Last May it issued its own energy policy, proconservation and antinuclear, which failed to pass the board in 1978 but which was jolted through following the trauma of Three Mile Island. The NCC has long been at odds with the Atomic Industrial Forum, with which it had what Carl Goldstein of the AIF calls a "long arduous debate" last year. Goldstein says the debates staged by the NCC have been "badly skewed" against all the conventional nonrenewable energy sources, and he feels members of the organization are "turning their backs on energy needs."

Church groups have made it clear that ecological justice, whether or not a "new theology" is developed, is now to be regarded as part of their mission. With the resurgence of religion in this country, that old-fashioned term "sin," which has been drowned in the currents of moral relativism, is making a comeback. A phrase in the NCC booklet on energy and ethics proclaims that humanity's "perversion of dominion into domination [of nature] is a sin and it is one of the underlying causes of the energy crisis." The new religious movements of the past decade have stressed exploration of the inner person. But as human dependence on the fragile biosphere becomes ever more apparent, movements of the 1980's may look more like neo-Pantheism.

Resources

Many of the written resources in this listing will suggest other resources you may wish to use. Almost all of the books and other publications listed here, as well as the written material accompanying the films, filmstrips, cassettes, and slides, list further reading, especially in the area of articles in magazines, journals, and newspapers.

Books
Books without acknowledged authors or editors are alphabetized by key word of title.

Brown, James, editor, *Time Bomb: A Nuclear Reader from The Progressive,* The Progressive Inc. (The Progressive Foundation, 315 W. Gorham St., Madison, Wisc. 53703), 1980.

Energy Stewardship: Energy Conservation Analysis of Three Massachusetts Churches, Total Environment Action, Inc. (Harrisville, N.H.), 1977.

Fenn, Scott, *The Nuclear Power Debate: Issues and Choices,* Investor Responsibility Research Center, Inc. (1522 K Street, N.W., Washington, D.C. 20005), 1980.

Hessel, Dieter T., editor, *Energy Ethics: A Christian Response,* Friendship Press, 1980.

Hoffman, Douglas R., editor, *The Energy-Efficient Church,* The Pilgrim Press, 1979.

In the Bank . . . Or Up the Chimmney?, U.S. Department of Housing and Urban Development, 1975.

Lea, William S., *Faith and Science: Mutual Responsibility for a Human Future,* Forward Movement Press.

Pimentel, David and Marcia, *Food, Energy and Society,* Halsted Press (Box 1313, Somerset, N.J. 08873), 1979.

Schumacher, E.F., *Small is Beautiful,* Harper and Row, N.Y., 1973.

The Solar Home Book; Heating, Cooling, and Designing with the Sun, Cheshire Books (Harrisville, N.H.), 1976.

Stobaugh, Robert, and Yergin, Daniel, *Future Energy: Report of the Energy Project at the Harvard Business School,* Random House, N.Y., 1979.

Other Publications: Pamphlets, Articles, and Resource Books

The Citizen's Energy Directory, Citizen's Energy Project, 1110 6th Street, N.W., Washington, D.C. 20003.

NCCC *Energy Packet,* NCC Energy Project, 475 Riverside Drive (Room 572), New York, N.Y. 10027. **In Canada,** order from Presbyterian Publications, 52 Wynford Drive, Don Mills, Ontario M3C 1/8, Canada.

Energy 80, The Christian Science Reprints Service, PO Box 527, Back Bay Station, Boston, MA 02117.

Community Energy Conservation Conferences: A Manual for Organizers, U.S. Department of Housing and Urban Development, Office of Neighborhoods, Voluntary Associations, and Consumer Protection, Washington, D.C.

Reducing Energy Costs in Religious Buildings, The Massachusetts Energy Office, Boston, Massachusetts.

Energy for My Neighbor: An Action Programme of the World Council of Churches, Department on Church and Society, World Council of Churches, 150 Route de Ferney, 1211 Geneva 20, Switzerland.

A Covenant Group for Lifestyle Assessment: Participant's Manual by William E. Gibson, Church Education Services, 1978, The Program Agency, The United Presbyterian Church, U.S.A., 475 Riverside Drive, New York, N.Y. 10027.

Consumer Reports: Buying Guide Issue, 1980 (Consumers Union of the United States, Inc., 256 Washington Street, Mount Vernon, N.Y. 10550.) "Energy Conservation."

Films, Filmstrips, and Cassettes

27 Energy Films (a catalog), The U.S. Department of Energy, National AudioVisual Center, Washington, D.C. 20409.

The Power To Change (a 28 minute, 16mm color film), Third Eye Films, 12 Arrow Street, Cambridge, MA 02138.

Holy Smoke: Biblical Reflections on the Energy Crisis (a 20-minute filmstrip or slide presentation with study guide). Packard Manse Media Project, Box 450, Stoughton, MA 02072.

Religion and Energy in the '80's (Cassette No. B5: keynote address by Dr. Elizabeth Bettenhausen, Assoc. Professor of Social Ethics and Theology, Boston University School of Theology, with selected questions from the Theological Imperatives discussion), NCC Cassettes, Room 860-R, 475 Riverside Drive, New York, N.Y. 10027.

170
Ces   R93-015

Cesaretti, C.A., ed.
THE PROMETHEUS
QUESTION

PROPERTY OF
ST. PAUL'S LIBRARY
9TH & GRACE STS.
PLEASE RETURN